Climate Change and the Future of Seattle

Anthem Environment and Sustainability Initiative

The **Anthem Environment and Sustainability Initiative (AESI)** seeks to push the frontiers of scholarship while simultaneously offering prescriptive and programmatic advice to policymakers and practitioners around the world. The programme publishes research monographs, professional and major reference works, upper-level textbooks and general interest titles. Professor Lawrence Susskind (MIT) acts as the General Editor of AESI, and oversees our book series, each featuring scholars, practitioners and business experts keen to link theory and practice. Our series editors include Brooke Hemming (US EPA), Shafiqul Islam (Tufts University), Saleem Ali (University of Delaware) and Richardson Dilworth (Center for Public Policy, Drexel University).

Strategies for Sustainable Development Series
Series Editor: Professor Lawrence Susskind (MIT)
Climate Change Science, Policy and Implementation
Series Editor: Dr. Brooke Hemming (US EPA)
Science Diplomacy: Managing Food, Energy and Water Sustainably
Series Editor: Professor Shafiqul Islam (Tufts University)
International Environmental Policy Series
Series Editor: Professor Saleem Ali (University of Delaware)
Big Data and Sustainable Cities Series
Climate Change and the Future of the North American City
Series Editor: Richardson Dilworth (Center for Public Policy, Drexel University, USA)

Included within the AESI is the Anthem EnviroExperts Review. Through this online micro-review site, Anthem Press seeks to build a community of practice involving scientists, policy analysts and activists committed to creating a clearer and deeper understanding of how ecological systems—at every level—operate, and how they have been damaged by unsustainable development. This site publishes short reviews of important books or reports in the environmental field, broadly defined. Visit the website: www.anthemenviroexperts.com.

Climate Change and the Future of Seattle

Yonn Dierwechter

ANTHEM PRESS

Anthem Press
An imprint of Wimbledon Publishing Company
www.anthempress.com

This edition first published in UK and USA 2022
by ANTHEM PRESS
75–76 Blackfriars Road, London SE1 8HA, UK
or PO Box 9779, London SW19 7ZG, UK
and
244 Madison Ave #116, New York, NY 10016, USA

First published in the UK and USA by Anthem Press in 2021

Copyright © Yonn Dierwechter 2022

The author asserts the moral right to be identified as the author of this work.

All rights reserved. Without limiting the rights under copyright reserved above, no part of this publication may be reproduced, stored or introduced into a retrieval system, or transmitted, in any form or by any means (electronic, mechanical, photocopying, recording or otherwise), without the prior written permission of both the copyright owner and the above publisher of this book.

British Library Cataloguing-in-Publication Data
A catalogue record for this book is available from the British Library.

Library of Congress Control Number: 2021931795

ISBN-13: 978-1-83998-545-4 (Pbk)
ISBN-10: 1-83998-545-3 (Pbk)

Cover Image: emperorcosar/Shutterstock.com

This title is also available as an e-book.

CONTENTS

List of Illustrations ix

List of Abbreviations xi

1. Introduction: Changing Seattle — 1
2. Background: Seattle's Green Development Story — 15
3. Current Situation: Building a "Climate-Friendly" City in an Unsustainable World — 35
4. The Future: Climate Change, Social Vulnerabilities, and the Transformational Agenda — 65
5. Conclusion: Seattle's Lessons — 91

References 99

Index 115

ILLUSTRATIONS

Box

2.1. Examples of Carbon Action Organizations in Seattle   30

Figures

1.1. Framing urban transformation as goals and issues   6
1.2. "Seattle" as a municipal, metropolitan, and relational space   9
2.1. Seattle's energy profile compared with US average   22
2.2. Shifting class structure in Seattle, 2010–17   25
3.1. Sustainable urban development and climate change   39
3.2. Territorializing carbon action: Net land use acreage in Seattle in 2016, both "inside" and "outside" the city's urban centers and villages   44
3.3. Areas of Seattle at "high risk" of social displacement   47
3.4. City of Seattle total budget expenditures, by area and percentage 53
4.1. Homeless encampments along the I-5 Corridor in Seattle (2018)   72
4.2. High Point's low-impact development, West Seattle   77
4.3. Integrating mitigation with adaptation in Seattle   80
5.1. Integration of water management with urban planning and urban design in suburban Seattle   95

Table

4.1. Climate Change Projections for Seattle, Washington   67

ABBREVIATIONS

ARC3.2	Second Assessment Report for Climate Change and Cities
CAP	climate action plan
CCA	climate change adaptation
CDC	community development corporation
CDP	Census Designated Places
CHOP	Capitol Hill Organized Protest
CNCA	Carbon Neutral Cities Alliance
CO2e	carbon dioxide equivalent gases
COP-3	Conference of Parties-3
DRR	disaster risk reduction
EPOC	Environmental Professionals of Color
FRED	Fund to Reduce Energy Demand
GCF	Global Climate Fund
GCoM	Global Covenant of Mayors
GHG	greenhouse gas
GMA	Growth Management Act
HOPE VI	Housing Opportunities for People Everywhere
HOV	high-occupancy vehicle
ICLEI	International Council for Local Environmental Initiatives
ICT	Internet and communication technologies
IPCC	International Panel on Climate Change
K4C	King County–Cities Climate Collaboration
LA21	Local Agenda 21
LEED	Leadership in Energy and Environmental Design
MCPA	Mayors Climate Protection Agreement
MSA	metropolitan statistical area
NIMBY	not in my backyard
OEM	Office of Emergency Management
OSE	Office of Sustainability and Environment
PSRC	Puget Sound Regional Council
RCM	regional climate model

SFH	single-family housing
SPU	Seattle Public Utilities
STBD	Seattle Transportation Benefits District
TMN	transnational municipal network
TOD	transit-oriented development
UCCRN	Urban Climate Change Research Network
UNEP	United Nations Environment Programme
UNFCC	United Nations Framework Convention on Climate Change

Chapter 1
INTRODUCTION: CHANGING SEATTLE

> In Seattle [...] we find, through a lot of luck, some good decisions, and the luxury of what used to be an abundance of natural margin for error, that discussions of sustainability and sustainable development are possible when making many of the choices we face. [...] Even here, however, old habits, myths, and the desire for stuff makes sustainability a hard sell.
>
> Gary Lawrence (1996)

This book explores a range of political, policy, and project efforts in Seattle, Washington, to mitigate and adapt to the formidable reality of global climate change. Developing a framework suggested originally by the Urban Climate Change Research Network (UCCRN), the book's core analysis considers both tantalizing progress and tangible problems in Seattle's climate action initiatives so far, particularly with respect to integrating carbon mitigation with adaptation; advancing climate action networks; cogenerating risk information; coordinating disaster risk reduction with climate change adaptation; and, most importantly, focusing on historically and geographically disadvantaged populations.

Linking together past, present, and future, I argue in what follows here that Seattle in the 2020s is less an "Emerald City" than an "Elite Emerald." Profoundly uncomfortable with this contradiction, local climate change efforts are shaped by mounting political concerns not only with mitigation-adaptation commitments and risk aversion policies to manage rising sea levels, warmer temperatures, and more variable rainfall patterns but also with reshaping a metropolitan space-economy that too often favors and consistently rewards the high-tech "cognitariat" over middle- and low-income households and communities of color.

Seattle's climate change efforts and urban challenges are, of course, part of a larger and more important global story. International concerns with the deterioration of the planet's ecological commons started to merge with work in urban affairs and local development strategies in the 1970s. This is best symbolized by the co-location, staffing, and funding in Nairobi, Kenya, of the

United Nations Environment Programme (UNEP) in 1972 and the United Nations Human Settlements Programme (UN-Habitat) in 1978 (Brand, 2005; Ivanova, 2010). The famous Brundtland Report accelerated the normative innovation of "sustainable development" in the mid-1980s. Part slogan, part program, this syncretic concept soon became a global aspiration that impacted policy practices around the environmental and equity dimensions of rapid urbanization along with ethical commitments to future generations (Næss, 1989).

The strategic role of cities in addressing global climate change is a more recent development, although it has been percolating for some time. The promulgation of Local Agenda 21 (LA21) programs after the Rio Earth Summit in 1992 elevated the strategic place of cities in global environmental politics more generally, while the Kyoto Protocol extension (COP-3) of the United Nations Framework Convention on Climate Change concomitantly helped to forge nascent linkages between LA21 efforts and global climate change concerns (Barrutia & Echebarria, 2011).

"International" programs like LA21 and, since 2013, the Sustainable Development Goals were (and are) less prominent in the United States. Still, the City of Seattle's (1994) landmark comprehensive plan, first published in 1994 and discussed in Chapter 3, provides an early example of urban-scale efforts within the United States to consider seriously the global impacts of "climate-altering greenhouse gases" (p. vii) and the attendant urgency of "carbon-neutral" (p. 2.37) urban development strategies. Strong evidence also suggests that the pace of policy changes and global carbon concerns in Seattle quickened substantially thereafter, especially as federal-scale climate policy faltered with the election in 2000 of George W. Bush (Office of Sustainability and Environment, 2002–2005).

At a conference held in Seattle in 1999, keynote speakers discussed "a battle for and against the Kyoto Protocol" taking place at the local level in the United States, "with governmental, business and environmental interest spending tens of millions of dollars to sway public opinion" (Miro & Cox, 1999, p. 25). In 2005, a partnership between the International Council for Local Environmental Initiatives (ICLEI), the Chicago Climate Exchange, the City of Seattle, Generation Earth, and two hundred other cities across the United States strengthened the Cities for Climate Protection program, targeting domestic sources of greenhouse gas emissions through new urban support for global climate protection. By 2006, Seattle had become one of the first big cities in the United States to adopt a climate action plan (City of Seattle, 2013, p. 2). Surrounding King County was the nation's first local government authority to require greenhouse gas auditing in its permit review system (King County, 2011). King County also spearheaded a new form of what I shall call later in this book "green city-regionalism" through the King County–Cities Climate Collaboration (K4C). Along with, this carbon reduction policy network consists today of the City of Seattle, the Port of Seattle, and 15 other suburban municipalities that collectively seek to coordinate and enhance the practical effectiveness of local government climate and sustainability actions.

In addition, a remarkable new "urban geopolitics" of carbon action has emerged (Bouteligier, 2012; Causone, 2018; Dalby, 2014). Major cities like London, New York, Copenhagen, San Jose, Barcelona, and Seattle are today not just key *sites of* but also key *actors in* global-scale efforts at climate mitigation and adaptation (Causone, 2018; Johnson, 2018). International relations have "environmentalized," in other words, even as global environmental politics have "urbanized" (Dierwechter, 2019). Leading transnational municipal networks like C40 Cities, United Cities and Local Governments, ICLEI—Cities for Sustainability, and the Global Covenant of Mayors exemplify this urban turn in world affairs (Acuto, 2013). In consequence, more coordinated policy action under the rubric of climate change and cities—or what the historian of global governance, Mark Mazower (2012), calls "scientific internationalism"—has returned to prominence. Barcelona, London, New York, Cape Town, Toronto, and Seattle—these and other cities are increasingly seen as "alternative global actors and problem solvers" (Haupt & Coppola, 2019, p. 123). From this perspective, cities like Seattle along with their local and transnational partners are the politically accessible places where science, participation, debate, and policy are creatively comingling to generate potential solutions and new approaches to climate change (Kristin, 2015). So hypothesized, cities like Seattle (especially in networks) are now the new spaces of green hope.

The Urban Climate Change Research Network and the *Second Assessment Report for Climate Change and Cities* Framework

Problem solving to expand these spaces of hope depends on political will and skilled leadership. Similarly, these depend on research and policy expertise. This book deploys the framework of the *Second Assessment Report for Climate Change and Cities* (ARC3.2) published in 2018 by the UCCRN. Associated originally with C40 Cities, the UCCRN is today one of several global-scale initiatives dedicated explicitly to the analysis of climate change mitigation and adaptation "from an urban perspective" (UCCRN, 2018). Analogous urban perspectives on global climate change are found in initiatives by the World Bank, OECD, and the UN-Habitat program. These initiatives, though distinct, frequently inform one another in practice.

One example illustrates this trend. Working originally with its UNEP neighbor as well as with Cities Alliance, ICLEI Climate Programme, the World Bank, and the Global Gender and Climate Alliance, the UN-Habitat's "Cities and Climate Change Initiative" has promoted, monitored, and reported on local implementations of the United Nations' "Habitat Agenda" since 2009. They have focused on questions of urban governance, capacity building, environmental planning, and mitigation and adaptation to climate change (UN-Habitat, 2009). As a "network of networks," the Cities and Climate Change Initiative has recently partnered with several transnational groups—including United Cities and Local Governments and C40 Cities—to integrate and formalize what a recent declaration signed in Edmonton, Canada, has called "The Science We Need for the Cities We Want" (for details, see https://c40-production-images.s3.amazonaws.com/other_uploads/images/1693_The-Science-We-need-for-Cities-We-want_March-7_Edmonton-1.original.pdf?1521742309).

One of the cosigners of that declaration, the UCCRN has explored the emerging role of cities in tackling global climate change, notably in its two major ARC3 reports. In doing so, the UCCRN has also helped to clarify "the science we need" even as it considers how we might actually "make," as the urban scholar Susan Fainstein (1999) originally wondered, "the cities we want" (assuming that we know "what we want" theoretically is in fact "what we need" empirically). For Fainstein and many other urban scholars, the core questions to ask are therefore not only what kinds of science we need—as important as these questions are—but also what kinds of political economy, especially when broaching distributional issues of equity and social justice in unequal, post-Keynesian societies (Piketty, 2014).

The two ARC3 reports, and particularly the second report (hereafter referred to simply as ARC3.2), identify various "pathways" that cities like Seattle can and should follow as they seek to "transform" their local economies, geographies, polities, and societies in order to ameliorate the profound vulnerabilities increasingly associated with what we know so far about global climate change (Rosenzweig et al., 2018).

I largely interpret these "pathways" as a series of *descriptive* rather than *causal* propositions as their theoretical interplay is not actually specified, nor is it clear from ARC3.2 why particular pathways emerge as they do in particular places beyond a generic sense of growing ecological crisis (see Zahran et al., 2008). I take up these points later in the book but simply note here that work in sustainability transitions might help to address these theoretical concerns and conceptual challenges (e.g., Geels, 20011).

In contrast to scholars like Clapp and Dauvergne (2008), who see contending "paths to a green world"—from "market-liberal" paths to more activist "social green" paths—the UCCRN (2018) instead sees, at least in broad conceptual terms, complementary pathways that draw upon a range of scholarly disciplines, ontological commitments, methods of research, and normative values (Figure 1.1). So framed, cities that are actively planning for climate change, ARC3.2 suggests, are vigorously following at least one (or hopefully more than one) of the following five "pathways" to urban transformation:

1. *Integrate mitigation and adaptation*, which are considered "win-win" actions that reduce greenhouse gas emissions while increasing resiliency;
2. *Coordinate disaster risk reduction and climate change adaption*, which are also considered the "cornerstones of resilient cities";
3. *Cogenerate risk information*, which assumes that involving "a full range of stake-holders and scientists" is the most effective approach;
4. *Focus on disadvantaged populations*, wherein "the needs of the most disadvantaged and vulnerable citizens" in a city are meaningfully addressed in plans and actions; and
5. *Advance governance, finance, and knowledge networks*, a process of multi-scalar networking that helps to build more robust city institutions, advances creditworthiness, and strengthens city research capabilities and policy expertise. (UCCRN, 2018, p. 4)

The theoretically ideal outcome in the ARC3.2 framework is therefore urban transformation—the loadstar condition—along these five distinctive pathways. This is visualized symbolically in Figure 1.1 as a "pentagon" of locally pertinent climate change programs and initiatives.

Obviously, real cities—places like Seattle—fall profoundly short of pentagonal transformation, whether conceived theoretically with mainstream economists as interlocking forms of Pareto optimality or with more heterodox ecological theorists as a "steady-state" economy (Daly, 1973). Neither exist anywhere except in textbooks. Cities likewise differ in their chosen areas of policy emphasis in everyday practice. For instance, "City A" might make more progress than "City B" in some areas, for whatever reasons, but not in others; alternatively, "City C" might pursue few, if any, pathways, while others might pursue many or even all of them. A globally uneven picture of policy change and urban transformation invariably emerges between (and within cities) and their metropolitan regions (Dierwechter, 2010). For example, Seattle arguably has done as much as any US municipality, save New York, to advance "networks," a theme taken up later in this book; but as the home to high-tech Amazon, Microsoft, and similar knowledge-based corporations, it has concomitantly struggled to include "disadvantaged populations," even as recent policy efforts explicitly address social justice (City of Seattle, 2013, p. 18; 2016a).

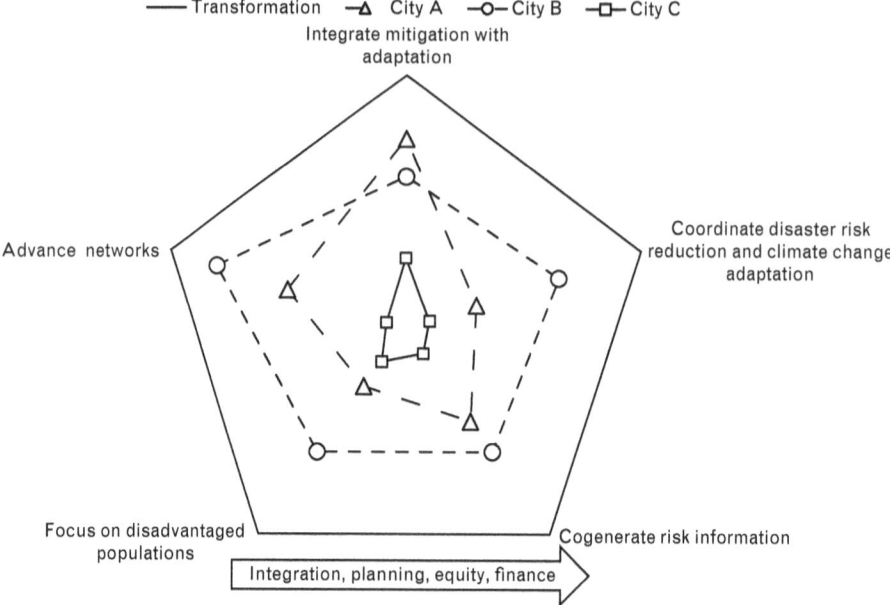

Figure 1.1. Framing urban transformation as goals and issues (Source: Author's rendering from Rosenweig et al. 2018, pp. xiii–4, 17–18)

Seen theoretically, contradictory patterns of urban development and social change are unsurprising. Equity-oriented urban policies suffer from the structural constraints of American political and legal systems, which tend to encourage zero-sum intercity competition for mobile (and now often global) capital at the long-term expense of social investments and redistributive policies (Schragger, 2013, p. 236). Yet nearly one-quarter of Seattle's populations of color live in poverty compared with only 9 percent of the Caucasian population (City of Seattle, 2016a). Moreover, the Seattle metropolitan region has not reduced *per capita* automobile emissions of CO2e (CO2 equivalent) as well as, for instance, metropolitan Portland, San Francisco, San Jose, or even Los Angeles (Gately et al., 2019). On the other hand, Seattle benefits enormously from extensive extra-urban hydropower infrastructure, which met 91 percent of the city's electricity demands in 2017; in contrast, the United States has remained disproportionately reliant on coal and natural gas (see http://www.seattle.gov/light/FuelMix/). Local realities and conditions matter a great deal, even as we must always consider these local realities and conditions in a multi-scalar framework.

"The Science We Need for the Cities We Want" thus begs many questions about the role of power as well as the conflicts and historical tensions that differently designed institutions of territorial and economic governance experience with one another in local political arenas (Dilworth, 2009). That acknowledged, the UCCRN's ARC3.2 framework provides a "multidisciplinary space" (p. 13) for various policy, practice, and scientific communities who are all committed to reimagining and rebuilding "post-carbon" cities that can eventually help to ameliorate global climate risks (Cleveland & Plastrik, 2018). Indeed, the ARC3.2 report pays recurrent conceptual and empirical attention to the world's most vulnerable urban populations—notably, disadvantaged citizens of the Global South whose daily struggle for basic survival in poorly serviced informal settlements and squatter areas is and will be made considerably harder by poorly managed global climate change (Dodman & Satterthwaite, 2008).

As Figure 1.1 shows, each pathway is tracked via "cross-cutting" themes. The organizing themes of the ARC3.2 study are (1) the core scientific and political capacity of any given city to integrate mitigation with adaptation by: (2) deploying new urban planning and design tools, (3) instituting equity and environmental justice policies, and not least (4) financing or leveraging sufficient resources from both public and private sector actors. Again, a crucial concern throughout is how these cross-cutting themes—simplified as integration, planning, equity, and finance—relate to one another within the empirical context of a given pathway, such as cogenerating risk information equitably or coordinating disaster risk reductions with adaptation efforts through new planning and design tools and financing strategies.

Assessing Seattle's Climate Traits *Relationally* within the ARC3.2 Framework

This brief book is a contribution to Anthem's series on *Climate Change and the Future of the North American City*. While broadly adopting the ARC3.2 framework of this series, the discussion that follows here considers the future of Seattle in light of present policies and approaches to climate change that have been shaped strongly by past institutional and socioeconomic histories and strategic policy priorities at multiple scales of governance and pragmatic action through various geographies of urbanism.

Seattle's climate action story only partly involves Seattle-placed actors. Many of the crucial challenges that Seattle faces have been and will long remain structured by wider dynamics. Some of these dynamics, such as older and ongoing state-level land-use planning reforms, arguably strengthen Seattle's "local" climate change efforts, but others, such as federal-level transportation and taxation policies, still fixated on building highways and subsidizing low-density residential development, concomitantly weaken these same local efforts. Accordingly, "Seattle" is understood in four analytical senses: as a political *place* (a bounded incorporated municipality); as a metropolitan statistical *area* (MSA); as an economic *web* of networked commuting and service flows; and finally, as a *metabolic source* of highly extended energy demands and provisioning grids, particularly as this involves water (Figure 1.2).

Taken together, Seattle is also a "glocalized" *space* as the planetary-wide challenge of urban climate action suggests; it is furthermore a metropolitan *complex* of people, nonhumans, and things as well as a multi-sectoral *bundling* of extra-urban networks and energy flows. The ARC3.2 report aptly calls the contemporary urban reality a "social-ecological system." Cities everywhere in the world, including Seattle, are "complex social-ecological systems (SES)," ARC3.2 suggests,

> uniquely endowed with attributes and functions that enable them to be the first and leading responders to climate change challenges in both mitigation and adaptation. [Such systems] are [therefore] dynamically interactive at multiple temporal and spatial scales. They consist of social and ecological components (broadly defined) that have their own internal processes; at the same time, these processes interact across the entire urban system. (p. 7)

In consequence, the ARC3.2 report specifically argues,

> As cities [like Seattle] mitigate the causes of climate change and adapt to new climate conditions, *profound changes* will be required *in urban energy, transportation, water resources, land use, ecosystems, growth patterns, consumption, and lifestyles*. New systems for urban sustainability will need to emerge that encompass *more cooperative* and integrated urban-rural, peri-urban, and *metropolitan regional linkages*. (p. 5, emphasis added)

Are we seeing any evidence of important changes in integration, planning, equity, and finance? Fitting Seattle into a multidimensional conceptualization of urban transformation—answering this synoptic research question—requires hard choices in a short book like this one about the most representative, interesting, and illuminating changes presently at work. It also means seeing Seattle concomitantly as an "urban" as well as a "city-regional" (or metropolitan-wide) "place" defined through variously constituted relationships. "Although we readily identify cities as the primary sites of adaptation planning and practice," Knieling (2016, p. x) usefully notes in his recent

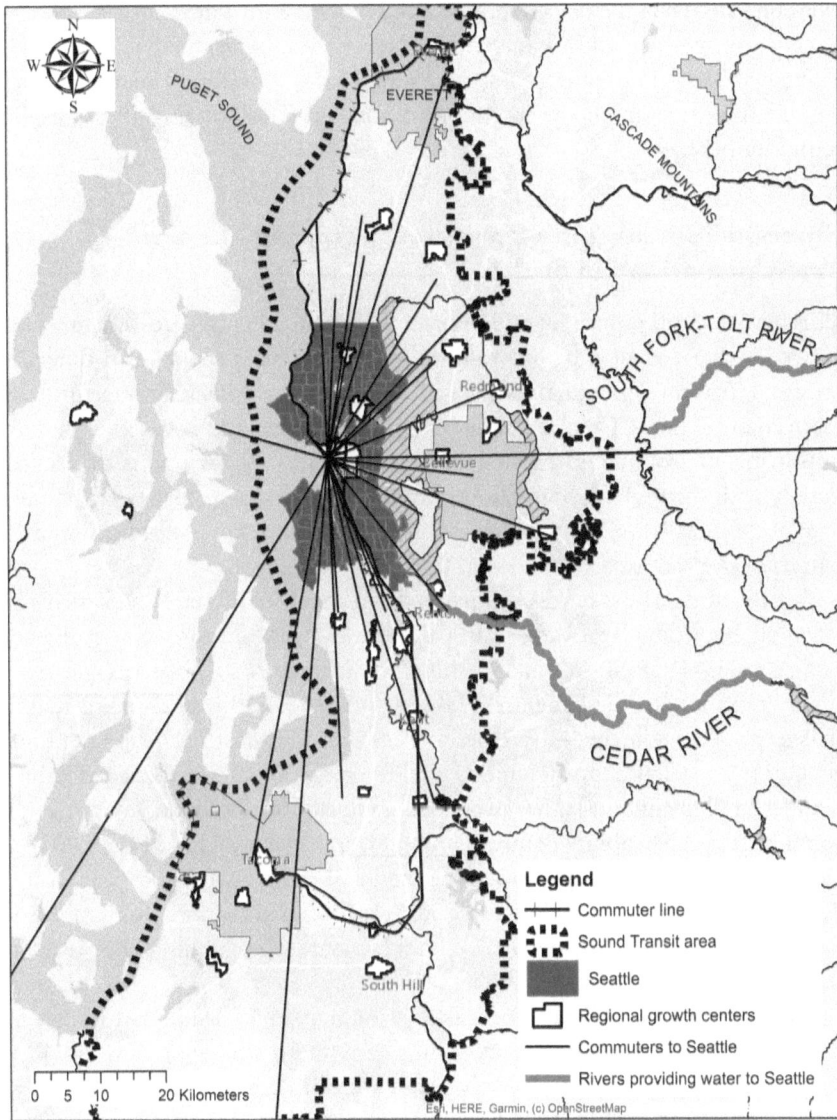

Figure 1.2. "Seattle" as a municipal, metropolitan, and relational space (Source: Author's rendering)

work on urban climate planning, "we also know that the nature of cities varies hugely around the world," which means that we might productively (and additionally) "focus on the metropolitan region rather than [only] on the entities within them that might or might not be called cities." I find this a useful reminder here.

Emerald Policies, Elite Pressures: Mapping "Urban Transformation" in Seattle

This book focuses especially on how Seattle is managing the growing tensions between what I shall call "emerald policies" and "elite pressures." Emerald polices refer here to enhancing local sustainability values, supporting climate change efforts, facilitating green spatial planning and urban design, and improving metropolitan–regional linkages. "Elite pressures" are associated mainly with the region's high-tech economic development story over the past several decades—and the attendant forms of race/class re-stratification and patterns of social exclusion (Fowler, 2016).

Seattle, of course, is hardly unique in this respect but arguably has lessons to teach others. Many large American cities—New York, Boston, San Francisco, Los Angeles, Denver, Austin, Washington, DC, and so on—struggle with similar tensions and challenges. As Himes (2019) notes, the demand for highly skilled workers has increased in cities that have developed strong agglomeration economies, accelerating wage inequality. In concrete policy practice, Seattle's planners, politicians, and other urban actors must actively negotiate a range of conflicts between livability, equity, economics, and ecology (Godschalk, 2004)—all while hopefully searching for a new model of *just* growth (Benner & Pastor, 2012), rather than just *growth*, which really only facilitates more rounds of "ecological gentrification" (Dooling, 2009) and "smart segregation" (Dierwechter, 2014). Is Seattle an emerald city—or is it an elite emerald?

No longer an industrial city or classic "company town" shaped by the once-dominant Boeing economy, Seattle is now routinely producing new kinds of socio-ecological spaces (Janos & McKendry, 2014). Some of Seattle's new spaces of change, such as the Duwamish River Basin, seek to "renature" the city's older industrial zones, which raises questions about the ecological and social effects of "post-industrial" transformation as well as the local distribution of benefits and burdens(Janos & McKendry, 2014; Klingle, 2007; Purcell, 2008). Still other spaces produced within Seattle and the wider metropolitan region seek to ameliorate the ecological and social impacts of urban development through reformed design standards (Vincent, 2019; Mathias, 2019), alternative transport investments, and novel land-use planning approaches (City of Seattle, 2016b; Dierwechter, 2014; Robinson et al., 2005). As a final example, new digital spaces seek to assemble and disperse "open data" in order to facilitate "algorithmic decision making" about local service provision and energy conservation—inaugurating so-called smart city capabilities that directly support global sustainability and climate action goals (AlAwadhi & Scholl, 2013).

Taken together, however, we need to ask: (how) are these new spaces actually "transformative"? (How) do they demonstrably generate "win-win" actions, reduce local disaster risks, increase urban resiliency, cogenerate critical information between stakeholders, meaningfully address disadvantaged populations, and/or advance new governance networks? (How) are these spaces empirically discernable in the region's urban energy grids, transportation infrastructures, water systems, land-use and land-growth patterns, and consumer behaviours?

The massive topic of climate change and cities is, once again, oriented strongly to thinking about likely and preferred urban futures. But as these questions each suggests, this future is based on the present and the present is shaped strongly by the past (or more accurately, multiple pasts) with different "temporalities," as ARC3.2 notes (ibid.). These different temporalities are still working themselves out today. Science-based policies must matter far more, but effective policies emerge from an institutional and cultural matrix of possibilities and capabilities that reflect antecedent choices and evolutionary path dependencies.

Plan for the Book

Within the context of these overall themes and questions, the remainder of this book is divided into four chapters.

Chapter 2 provides a retrospective look at Seattle's long-term development story within the analytical context of its wider ecological and metropolitan settings. Key topics include its origins as a frontier city and its structural relationship with California and the Yukon; its major periods of industrial change, demographic growth, and territorial expansion, especially the impacts of Boeing, Microsoft, and Amazon; its history of "boom-and-bust" cycles; and the parallel establishment of Seattle's primary infrastructural systems (parks, water and sewer, energy, public transit, highways, ship canals, bridges, etc.). The chapter further provides an overall demographic profile of Seattle, focusing on its current social and economic structure, including median family income, home ownership rates, home value changes, and the city's changing racial, ethnic and class makeup. These topics are relevant to the relationship between equity and the impacts of climate change, which again constitutes the book's organizing theme.

The discussion next shifts to a brief elaboration of Seattle's basic political and governance system, including: (1) the institutional structure of the city, notably the extent to which major infrastructure elements are controlled by the city and special-purpose governments (port authority, Metro, etc.); (2) the strong mayor form of the city's government; (3) the partisan makeup of the city and related issues (e.g., ideological continuity across mayoral administrations that support long-term climate action and planning); (4) the role of nongovernmental organizations in city governance; as well as (5) the legal and policy relationships between the municipality and Washington state agencies and regulations. Finally, the chapter concludes with a discussion of the city's primary climate change challenges and a special review of the key infrastructure and urban service systems that arguably face the greatest threats from climate change.

Chapter 3 narrows the analysis to engage with the recent story of city climate change policies and related urban sustainability plans, mainly starting from the early 1990s. The discussion explores the content of key plans, including mitigation and adaptation efforts in policy arenas like urban infrastructure, the built environment, urban form, transport, and carbon sequestration. The chapter also elaborates the structure and significance of city officials whose jobs have been—and are—related to climate change action. Seattle actively participates in both national and international climate action networks, including the Mayors Climate Protection Agreement, C40 Cities, the Global Covenant of Mayors, and Carbon Neutral Cities Alliance, a theme explored with reference to key policy arenas (e.g., zoning and planning, transportation, drainage). Within this context, the chapter considers the leadership role of officials and agencies as well as relationships with the Port of Seattle; county and regional transportation authorities; and nearby municipalities like Bellevue, Redmond, and Renton, King County; and the Puget Sound Regional Council. Finally, the chapter addresses the city's relationships with nonprofits, foundations, and industry associations as well as state-level environmental authorities and, where relevant, federal institutions.

Using the UCCRN projections from ARC3 and ARC3.2 as well as research conducted by local climate change experts over the past 20 years, Chapter 4 first speculates on the most likely scenario for Seattle in 2100. Imagining a future climate setting possibly more like contemporary San Francisco—more rain overall but with wetter winters, drier summers, rising temperatures, and higher sea levels—the chapter focuses on questions such as what the city and its partners have been doing to prepare for such a dramatic transformation and how likely (or not) these actors will be able to meet the local challenges of global climate change. In particular, the chapter explores the city's assets and obstacles to change as well as the risks of insufficient policy action to confront the most important obstacles. The discussion here draws empirically on the insights of local scientists and specialists in urban climate action and advocacy and places key arguments within the ARC3.2 conceptual framework on climate-proofing major cities and urban regions in North America and beyond.

As a conclusion, Chapter 5 recapitulates and synthesizes the insights and lessons that the city-region of Seattle provides to students of urban climate change, carbon policy, and strategic planning within and beyond the American and North American context. The chapter identifies the key ideas and themes—and lessons—from the previous four chapters and relates them to the wider themes, concerns, and recommendations advanced by the UCCRN and the ARC3.2 book, *Climate Change and Cities*. The most important of these is what kind of future(s) we might anticipate for Seattle and the metropolitan region.

Chapter 2

BACKGROUND: SEATTLE'S GREEN DEVELOPMENT STORY

> We need a new ethic of place, one that has room for salmon and skyscrapers, suburbs and wilderness, Mount Rainier and the Space Needle, one grounded in history.
>
> Matt Klingle (2007)

The city of Seattle and surrounding metropolitan communities—nearby cities, suburbs, towns, Indigenous nations, port authorities, regional transit providers, and counties—are collectively nestled along the comparatively deep, well-mixed fjordal estuaries and bays of Puget Sound within the Salish Sea of the Pacific Northwest. Built across re-engineered hills beneath majestic mountains, Seattle traverses bodies of both salt and fresh water. It draws on multiple river basins for sustenance. The wider region's mild marine air environment has encouraged prolific vegetation and abundant natural resources. These include the seemingly endless cedar trees and beaver pelts that first attracted European interests in the 1830s and then, in the 1850s, permanent American colonizers, initially from the Midwest. Like Cape Town, Rio, or Vancouver, Seattle is physically stunning—known in popular culture as "The Emerald City" for its lush rain forest canopies, wet-green hues, and relatively mild climate.

How did Seattle get here—and how do its various pasts still shape where it is going today or might go tomorrow? The broad outlines of Seattle's overall "development story," about which more below, are generally well known. That said, synoptic, synthetic, well-theorized accounts of Seattle's longer-running historical geography and key urbanization patterns are rare, especially when "Seattle" is embedded within its wider "post-metropolitan," institutional, and ecological contexts (Dierwechter, 2017; Sale, 1976).

My central argument in this chapter is that policy patterns associated with contemporary Seattle's post–carbon development choices remain shaped by "multiple orders" of development (Dierwechter, 2017), a term derived from the work of Orren and Skowronek (2004) in historic institutionalism. The larger goal is to help situate contemporary development problems and carbon policy patterns explicated in Chapters 3 and 4, where the ARC3.2 framework's emphasis on the multiple "pathways" of urban transformation in American cities is further engaged.

As stated in Chapter 1, the ARC3.2 framework advanced by UCCRN treats cities like Seattle as "complex socio-natural systems" whose constitutive components—physical, social, natural, and virtual—are "dynamically interactive at multiple temporal and spatial scales" yet also "have their own internal processes" (p. 7). Taking these big ideas seriously means, in my judgment, teasing out how "internal" processes "interact" over institutional time and across policy space, and no less how these interactions might eventually help to forge the more "cooperative arrangements" in regions that *will be required* to advance urban transformation through science-shaped agency. To go forward, then, we first go back—exploring where Seattle came from and hence where it currently stands.

Origins, Development Patterns, and Contemporary Demographics

What is today Seattle emerged in 1851, about half a century after the Lewis and Clark Expedition opened the Pacific Northwest Territory to an expanding mercantile political economy. It was a largely white society of colonial settlers led by figures like Arthur Denny, Carson Boren, Doc Maynard, and Henry Yessler, who disrupted and then steadily remade extant patterns of development that were long associated with the Salish cultures and Lushootseed-speaking peoples of the region: notably, the Suquamish, Duwamish, Nisqually, Snoqualmie, and Muckleshoot. Seattle was platted quickly in 1853 and then incorporated formally in 1865, effacing the village of Duwamps. The place name "Seattle" derives ironically from the name of the Duwamish chief, *Sealth*, who befriended Maynard and other pioneers even as he mythically warned of the new economy they had introduced.

The pioneer's new economy—capitalist, urban, proto-industrial—started in Seattle with Henry Yessler's steam-powered sawmill. Yesler's mill transformed the temperate rain forests nearby into board feet of refined lumber, not just to meet local construction demands but, in time, for export. Booming from the 49ers' Gold Rush and the economic windfalls of the Mexican-American War, San Francisco provided a major market for Seattle's early growth. Other emerging communities in the region, notably Tacoma, also provided lumber for export. But Tacoma's economy did not sufficiently diversify away from lumber and wood-related industries (McKean, 1941), whereas Seattle's development story reflected a clearer capacity to spin off new industrial competencies (MacDonald, 1987) or what Jane Jacobs (1969) later famously identified as "making new work from old work" (see also Abbott, 1992).

A second gold rush in the Klondike River–Yukon–Alaska territory in the late 1890s similarly stimulated Seattle's initial growth, solidifying new sectors in equipment provisioning, banking, insurance, and finance as well as in other commodity production lines and transport logistics. The Moran Brothers Co., for example, built a dozen paddlewheel boats for Yukon prospectors—and then the battleship *Nebraska* in 1904. "The Klondike Gold Rush," Montgomery and Mighetto (1988, p. 3) have argued, "fueled a longstanding commercial spirt that continues to this day," in part by focusing the boosterism of the Seattle Chamber of Commerce, which in turn encouraged a culture of risk-taking (Sale, 1976, p. 51). Yet as Blair (2014) argues, Seattle's early path dependencies were shaped by "fire" no less than "gold." Seattle was essentially, in Klingle's (2007, p. 54) apt words, "a transmuted forest and often burned like one." The devastating Seattle fire of 1889 paradoxically forced Seattle's businesses and public officials to agree on "a complete demolition and redesign of the entire business district" (Blair, 2014, p. 7). This crisis produced both immediate and long-term advantages (Ochsner & Andersen, 2002); specifically, the fire likely deepened local civic capital and the movement for municipal ownership of major utilities. Seattle's vulnerabilities and risks had their benefits—an experience not lost on present-day environmental activists and climate change policymakers.

By World War I, Seattle's population had reached 250,000, far surpassing that of its would-be rivals in the Pacific Northwest (Tacoma, Olympia, and Portland). By Armistice Day a few years later, another local entrepreneur was literally putting "the transmuted forest" into the air. Made originally out of wood, the new airplanes produced by William Boeing's new aeronautics firm did more than any other institution of any kind to shape Seattle's development story—at least until the flowering of Microsoft in the 1980s and the "digital" shift that would help to incubate another company, Amazon, in the mid-1990s.

Established on the Duwamish River in 1916, the "Boeing Airplane Co." eventually imprinted upon the economic region a "hub-and-spoke" model of industrial production, concentered originally along the Duwamish River corridor of manufacturing but soon radiating out across King, Snohomish, and Pierce Counties. As Gray et al. (1996, p. 651) explain, hub-and-spoke economies tend to cluster around large core firms just like Boeing. They often generate high regional growth rates and good income distributions—one reason why some observers think Seattle can avoid San Francisco's fate (Luis, 2012). But hub-and-spoke economies are famously vulnerable to cyclical patterns.

Born during World War I, Boeing long benefited from wartime spending as much as from peace and prosperity, suffering major downturns in revenue and employment after World War II and the inevitable dénouements of the Korean and Vietnam wars. Like Denver, Seattle's development story reflects a "boom-and-bust" rhythm; like Los Angeles, Seattle remains an important member of what Kirkendall (1994) once memorably called the country's "metropolitan-military-industrial complex" (p. 137). Certainly, Seattle is less shaped by Boeing than it once was. But Boeing's decision to move its corporate headquarters to Chicago in 2007 and, in 2020, consolidate 787 Dreamliner assembly in South Carolina highlights the city's economic vulnerabilities and regional concerns.

A shrinking city still reliant on Boeing in the 1970s, located "in the mists" far away from the nation's core markets (Luis, 2012), Seattle nonetheless built up new economic competences and value chains in the latter part of the twentieth century in desktop software, gaming, life sciences, cloud computing, interactive media, cyber security, and internet commerce (Prosperity Partnership, 2012)—taking evolutionary advantage of existing pools of high-tech labor, a perceived high quality of life, and a major research university with substantial federal grant activity (a key factor missing in neighboring Portland, for instance). Dipping to under 500,000 people in 1980, Seattle reached 750,000 in 2020. Mayer (2013) shows in her work on "spin-off firms" across the Seattle city-region that Microsoft facilitated the *in situ* creation of an "entrepreneurial ecology" characterized by "knowledge spillovers" and "regional branching," which helped the metropolitan area not only to grow but to diversify over time (p. 1711).

Undoubtedly so. But the maturation of Amazon, founded near Seattle in 1994, reveals how powerful firms—global behemoths—can structure space and society. Rivaling the Great Fire of 1889, Amazon's recent successes have effectively redeveloped a significant part of Seattle's urban core for high-tech workers—sending shock waves across the city and region, with implications for housing affordability, gentrification patterns, race and class segregation, transportation systems, and, not least, traditional retail spaces like big-box stores and shopping malls. At only 1 percent of total retail sales in the United States in 2000, online shopping topped 16 percent in 2019—a figure expected to grow substantially in future years, with implications for labor politics, household income, warehousing complexes, and ultimately for patterns of carbon use. (Seattle's iconic downtown Macy's store has closed, as one example, with Amazon occupying its top floors.)

Processes of (re)development like these do not occur in a political vacuum. In addition to nature, they are shaped constantly by the *impress* of state regulation and public investment, particularly around the financing, provisioning, and management of bulk infrastructure systems that necessarily underpin all regimes of accumulation, for example, parks, water and sewer, energy, public transit, highways, ports, ship canals, and bridges. In this sense, it is misleading to describe Seattle's long-running political culture as "classically liberal," meaning a city and region mimetically reflecting a singular American culture that values "a government by consent, limited by the rule of law protecting individual rights, and a market economy" (Gerring, 2003, p. 92). That is true, but insufficient; it is also aspatial, as if America is a liberal "container" rather than a product of diverse territorial forces and unevenly expressed values and interests. In fact, the labor historian James Gregory (2015) argues that contemporary Seattle exhibits a "dual personality." Boomtown Seattle "has been re-engineered by billionaires," he observes, "but boomtown Seattle is also a progressive city, with loud echoes of a more radical past" (p. x).

Unleashing free markets on ecosystems and societies may well produce economic "booms," but this is not "progressive" much less "radical"—suggesting perhaps three (rather than two) traditions within Seattle's political history and institutional development (Dierwechter, 2017, chapter 5). Gold, airplanes, software, wars, online services—these are the "boom" agents, and they obviously have been critical. But progress reflects more fundamentally how the development process is shared and with whom, how it is reshaped to make complete communities, and how sensitively various natures are protected. Progressive cities nurture democracy and reinvest wealth for a wider social good, partly by "building the public city" through critical infrastructure systems that cannot be captured by elites (Perry, 1995).

Let us start with water systems. By 1870, just five years after incorporation, Seattle was already experienced enough with fire to form the Seattle Hook and Ladder Co. No. 1; that same year it passed an ordinance that required indoor plumbing at every dwelling. Water management expanded in 1882 after the city incorporated the Spring Hill Water Company; immediately after the Great Fire in 1889, though, voters created a municipal water system—approving revenue bonds to construct the Cedar River water system several miles south of the city. In 1891, the city acquired a key water-pump station through municipal ordinance and improved its sewerage and storm water systems—finally shedding its heretofore ad hoc and uncoordinated reliance on individual wells, springs, and private water companies. To meet the needs of a rapidly growing city, water system capacity grew astonishingly from 23.5 million gallons per day in 1901 to 68.5 million gallons per day in 1909. In 1964, additional capacity was added from the South Fork of the Tolt River, serving north Seattle and Eastside municipalities. Seattle's contemporary water system supplies about 180 million gallons of water per day to 1.4 million people across the Seattle city-region; about 70 percent is from the

Cedar River water system, which is anchored by a treatment facility in nearby Renton built in 2004 on Lake Washington.

Early infrastructure developments helped to leverage Seattle's annexation of nearby Wallingford, Green Lake, and the University District (today some of the city's trendiest areas) and, just a few years later, communities like Ballard, Georgetown, Rainer Beach, West Seattle, and Laurelhurst. Seattle would annex smaller portions of land in the 1940s and 1950s (and as late as the 1980s), but the city's legal footprint—its basic overall shape—was largely in place a hundred years ago. After this time, as was common across the United States, a series of suburban incorporations accelerated political, regulatory, and service fragmentation. Such fragmentation reflected a strong bias toward local control in American political culture—but it worked against a multitude of problems that required regional planning and metropolitan coordination.

This was especially clear for sewerage and storm water policy. While public control over both systems was established early in the history of Seattle, such "systems" were rudimentary, consisting of gravity-driven wooden boxes (i.e., "transmuted forests" below ground). Waste was untreated and simply funneled into near-shore waters, especially Elliott Bay, Lake Union, the Duwamish River, and Lake Washington. "Combined" rather than "separate" systems within the city expanded from 14.9 miles in 1891 to over 1,000 miles by 1956. By then, however, postwar growth and suburbanization finally reached a tipping point. With new incorporations, dozens of new sewerage districts had formed, although many unincorporated areas—a third of the region's metropolitan population—still lacked public sewerage services. Lake Washington was especially affected by the attendant pollution (King County, 1958, pp. 12–16).

"Cleaning up Lake Washington," and specifically the cyanobacteria *Oscillatoria rubescens*, became a rallying cry in the late 1950's push for more significant proposals that sought comprehensive reforms in the region's governance structures. The rallying cry worked. Between 1963 and 1968, local officials steered $140 million into new trunk lines and interceptors—the costliest pollution control program in the country at that time. Effluent to Lake Washington stopped in 1968, with recovery thereafter. Moreover, the region's voters created a new political body, the Metropolitan Municipality of Seattle, albeit with lukewarm suburban support. Originally conceived as a federated city-county agency to provide waste management, mass transit, regional parks, and regional planning services, voters only agreed to the waste management remit. The Metropolitan Municipality of Seattle took over regional transit in 1972 and was merged formally with King County in 1994. Seattle certainly tried. But it would never get the institutional and territorial equivalent of Portland's more famous Metro authority, with its directly elected regional councilors and greater planning and regulatory powers.

Like nearly everywhere in the United States, postwar and mid-century Seattle and the wider city-region were structured more by private cars than by public transit. The Washington State Highway Department rejected a proposal in 1953 for including rail transit options in what was then called the "Central Seattle Freeway" and today Interstate 5. This policy decision echoed the verdict of King County voters in 1952. They vetoed as "communist" another proposal to build a rapid transit system. Until the 1990s, political failures to invest in public transit systems overwhelmed smaller successes, such as the downtown "Magic Carpet" (or ride-free) zone, high-occupancy vehicle (HOV) lanes, and a "shared-commute" program of county-owned vans. The politics were always tough. In several cases, state law required "supermajorities" to move forward, whereas federal pro-highway policies only required simple majorities.

Yet as Seattle's regional economy shifted in the 1980s, new opportunities for public transit shifted as well, aided by major reforms in state-level planning and federal transit policies (notably, the state's landmark 1990/1991 Growth Management Act). A new regional transit agency created in 1993, Sound Transit, steadily integrated investment practices across the entire Seattle–Tacoma–Bellevue metropolitan statistical area (MSA), especially for express bus and light and commuter rail systems. Sound Transit today sits alongside of several other public transit agencies, including Pierce Transit, Community Transit, the state ferry system, and King County Metro—a governance approach bemoaned by many public officials. That said, voters in the Sound Transit district approved three major referenda (1996, 2008, and 2016) to fund the agency's transit investment plans through 2041. One challenge here has been—and remains—how to coordinate regional plans for light and commuter rail systems with local land-use priorities, especially around transit-oriented developments (TODs) that address ecological and social justice concerns no less than commuting needs (Bakkente et al., 2015; Lowe, 2014; Bakkente, 2020). As we shall see in later chapters, changing metropolitan transport geographies away from private automobility is a crucial aspect of climate policies (Raven et al., 2018).

No less crucial is how electricity is generated (Davoudi et al., 2009). This underscores the history and future of public utility grids as "socio-technical systems" that, in the present case, extend beyond urban space. At about the same time that William Boeing was reinventing aircraft production, the Seattle Board of Public Works in 1918 was designing another hydroelectric plant to accommodate the voracious energy demands of the city's burgeoning population—complementing its first hydroelectricity project on the self-same Cedar River in 1905, the first municipally owned hydro project in the country (Crowley et al., 2010). A rainy environment had its positive effects, too, even as

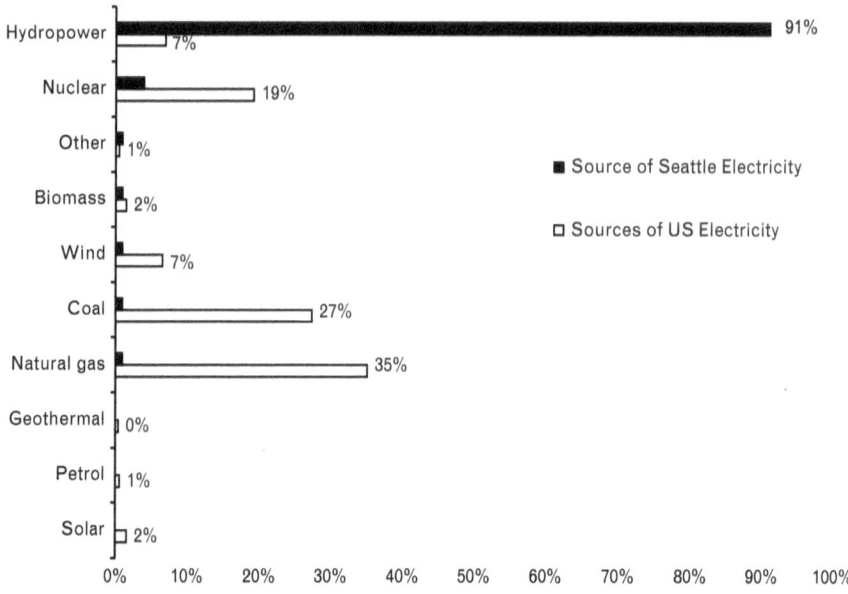

Figure 2.1. Seattle's energy profile compared with US average (Source: Author, data based on seattle.gov/light/fuelmix/)

ideological struggles between private and public providers at that time shaped the institutional landscape of early energy systems—with hostility to natural monopolies pitted against concerns with the equity implications of market-based energy supply (ibid.). Seattle City Light formed officially in 1910. In subsequent decades, it built new dams, storage facilities, grids, and even rails to deliver "green" hydropower to the city, initially along the Skagit River. By 1951, all remaining private electricity vendors were bought out.

As shown in Figure 2.1, 91 percent of Seattle's electricity is generated by hydropower, far outpacing the average of only 7 percent for the United States. While this is impressive at face value, cities like Reykjavik, Iceland, source 100 percent of their electricity from hydropower and geothermal (as does the much smaller Burlington, Vermont). Today Reykjavik further seeks to power cars and public transit with fossil-free energy. Greener outcomes are still needed, not only within but across the major socio-ecological systems of all world cities (UN Framework Convention on Climate Change, 2018).

Creating effective, real-time, synergistic linkages across ordinary urban services—water, energy, sewerage, transit, built environmental, and so on—constitutes the most recent "infrastructural" challenge that today's cities now face, Reykjavik, Burlington, and Seattle included. Such linkages represent "smart transitions" in territorial governance and accumulation regimes, with major implications over time for city-regional competitiveness and sustainability policies (Herrschel & Dierwechter, 2018). Nonetheless, whether such transitions are sufficiently comprehensive is in question.

Central to "transitions" are building "smart city" capabilities, which refer to digital infrastructures that "augment" urban spaces and systems through a meshwork of sensors, apps, actuators, dashboards, Big Data control enters, and/or other modes of real-time data harvesting and algorithmic decision-making around urban service provision (Coletta et al., 2019). Greenfield smart cities like Masdar City, PlanIT Valley, or Songdo have received the most attention. But more relevant here is how extant cities with extant socio-technical systems are now weaving new digital capabilities into infrastructure portfolios and governance routines (Glasmeier & Christopherson, 2015; Shelton et al., 2015). In theory, these new digital infrastructures can overlap with efforts to reduce greenhouse gas emissions through, for instance, the management of zero-energy buildings (Kylili & Fokaides, 2015). As one of the county's "tech hubs," Seattle would appear well positioned to innovate further in this area of urban infrastructure (Soper, 2017).

Already by 2012, AlAwadhi and Scholl (2013) had located 20 smart city initiatives in Seattle. Many were (and remain) standard projects that we now see everywhere, for example, open data portals, the Electronic Plan Review System, the Digital Evidence Management System, new automated metering infrastructures, a smart grid, and augmented drainage and wastewater. Other initiatives were more unexpected, such as the Equitable Justice Delivery System. That system has deployed internet and communication technologies (ICT) infrastructure to help advance economic mobility and opportunity; prevent residential, commercial, and cultural displacement; and promote transportation mobility and connectivity, among other goals. Comparing these initiatives with Philadelphia, Mexico City, and Montreal, though, AlAwadhi and Scholl (2013) found that Seattle was, at that time, advanced in this policy arena (and see Seattle's early progress compared with that of Vancouver, Canada (City of Vancouver, 2013)).

A major question is whether, how, and where such smart city efforts in Seattle help with the larger crisis of climate change mitigation, adaption, and local resiliency. Critiques of smart city investments in Seattle as elsewhere are increasingly common (Scott, 2016). Coletta et al. (2019) thus summarize the current state of the smart city debate as an "impasse":

> On the one side are those that seek to develop and implement smart city technologies and initiatives, often with little or no critical reflection on

how they fit into and reproduce a particular form of political economy and their wider consequences beyond their desired effects (such as improving efficiency, productivity, competitiveness, sustainability, resilience, safety, security, etc.). [...] On the other side, the smart city [...] facilitates and produces instrumental, functionalist, technocratic, top-down forms of governance and government [...]; is underpinned by an ethos of stewardship (for citizens) or civic paternalism (what is best for citizens) rather than involving active citizen participation in addressing local issues. (p. x)

Arguably, both propositions are true. Cities like Seattle are formed through contradictions from multiple orders of development and policy change, a theme Robert Beauregard (2018) has applied recently to urban sustainability. Some orders reproduce a technocratic, top-down "boom city"; other orders progressively tackle "root and structural causes." The aforementioned Equitable Justice Delivery System in Seattle, for instance, has sought over the entire course of the 2010s to "embed race and social justice and service equity" across Seattle's public utility services, placing "environmental justice" at the core of the city's wider equity goals. Seattle's public utilities staff use an Equity Planning Toolkit in their everyday work to engage in outreach activities, even as a sophisticated new grid of sensors and actuators seek algorithmic efficiencies with long-term climate mitigation benefits (Seattle Public Utilities, 2019).

Seattle's changing demographics and its social and economic structure warrant this growing concern with equity and inclusion.[1] Approaching a population of 750,000 in 2020, Seattle has been one of the country's fastest-growing large cities in recent years. Only the cities of Phoenix and San Antonio have grown faster. When stretched functionally to include King County, the Seattle region has been the third fastest-growing economy in the United States over the past 10 years. Out of 384 MSAs, moreover, the population of Seattle–Bellevue–Tacoma MSA has grown 15 percent since 2010, placing it just behind the Phoenix, Dallas, and Houston MSAs. In 2018, Seattle was the 18th largest city in the United States; the Seattle–Tacoma–Bellevue

---

[1] Unless otherwise indicated, data reported in the next few paragraphs on the social and economy structure of Seattle and the Seattle–Bellevue–Tacoma MSA are gleaned from the following sources: the US Census Bureau factfinder tool and the US Bureau of Economic Analysis, GDP, and personnel income. See US Census Bureau: https://factfinder.census.gov/faces/nav/jsf/pages/index.xhtml; US Bureau of Economic Analysis: https://apps.bea.gov/itable/iTable.cfm?ReqID=70&step=1.

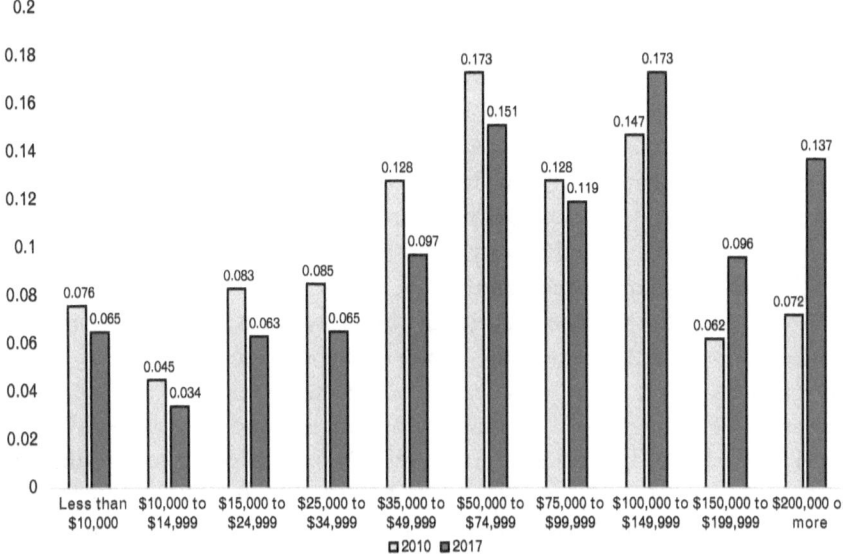

Figure 2.2. Shifting class structure in Seattle, 2010–17 (Source: Author, based on US census data)

MSA was the 11th largest regional economy, and since 2010 the county's third fastest-growing economy, ahead of Atlanta, and just behind the San Francisco and San Jose MSAs.

As might be expected, this dynamic growth has significantly remade the region's geographies of economic benefits and social burdens. Over the past decade, for example, the class structure within Seattle has discernably changed. Seattle exemplifies the post-Fordist divergence of household wages both between and within places, amplifying the deleterious effects of what Dreier, Mollenkopf, and Swanstrom (2014) have called the economic segregation of the United States. The tech-driven industrial clustering (and regional branching) around massive firms like Microsoft and Amazon discussed earlier has mostly benefited high-skilled workers. As seen in Figure 2.2, more and more of Seattle's overall population is made up of well-educated households earning $100,000 year or more—and especially households earning $150,000–200,000 or more. At the same time, middle-class households make up less and less of Seattle's overall social structure.

Wealth dynamics like these have immediate impacts on rates of home ownership and attendant home values. Like most central cities in the United States, about half of Seattle's 334,850 housing units in 2017 were rentals. Around 43 percent of housing options in Seattle that same year were single-family detached units; another 32 percent were in tower developments with 20 or more units (i.e., condos). Duplexes, fourplexes, and other types of clustered housing were uncommon (Dierwechter, 2014)—even as scholars associate these "missing middle" forms of housing with equity policies and walkable urbanism, both nominally key goals in the city's sustainable urban development strategies (Parolek, 2016). In part, this is a political product of how increasingly wealthy residents maintain barriers to code reforms, inclusionary zoning tools, and targeted density bonuses that, at least in planning theory, facilitate housing choice and equity goals.

Long called the defense of privilege by growth scholars, such territorial politics of regulation reflect the dramatic appreciation of median home values in Seattle in recent decades—as well as what many critical observers see as the pathologies of sometimes subtle racism and class biases (Lyons, 2004, pp. 64–84). In mid-2019, the median home price in Seattle, though cooling off, was $714,600 compared with $489,400 for the MSA as a whole—appreciating about *six* times faster than individual incomes over the past several years (Long, 2019). Among the country's hundred largest MSAs, the Seattle–Tacoma–Bellevue MSA was ranked the fifth least affordable region in 2019, outpaced only by the San Jose, San Francisco, Los Angeles, Honolulu, and Oxnard, California MSAs (Clear Capital, 2019). Unsurprisingly, homelessness numbers have skyrocketed, not only in Seattle but, over the latter 2010s, in nearby Tacoma. About 40 percent of Seattle's homeless population of 12,000 in 2018 was African American.

Far away from both the American South and Mexico, Seattle remains unusually white for a large American city, although the booming Boeing economy attracted major African American migration in the 1940s especially (Taylor, 1994). About 68 percent of Seattle's population in 2017 was Caucasian, compared with, for instance, 53 percent in Boston and 30 percent in Baltimore. Seattle's Asian (especially Japanese, Chinese, Indian) and Latin American populations are growing in both absolute and relative terms, but the Black and African American populations, with roots in the nineteenth century, are shrinking as a proportion of the city's overall population (down to only 7 percent). The historic Central District in Seattle, once home to Jimmy Hendrix and Quincy Jones, was 80 percent African American in the 1970s and 75 percent of the city's entire African American population. By 2010, it was majority Caucasian, as the gentrification effects of the city's residential sorting and condo-building cascaded down from Capitol Hill, Ballard, and Freemont—extruding many African American families to Rainier Beach and also to cosmopolitan suburbs like SeaTac (Balk, 2014).

Review of the Political and Governance System

In 1878, voters in Seattle elected as their racist mayor the "People's Ticket" candidate, Beriah Brown, who nonetheless championed civil rights for African Americans in the still-young University of Washington (Tate, 2000). In 1941, the Seattle Housing Authority built Yesler Terrace, the first racially integrated public housing development in the United States. In 1990, Norm Rice was elected Seattle's first black mayor. In 2004, Seattle's mayor, Greg Nickels, launched the Mayors Climate Protection Agreement, openly defying the Bush administration's infamous rejection of the COP-3 Kyoto Climate Accords, just a decade after Seattle published the first comprehensive plan built explicitly around urban sustainability values. In 2016, over 80 percent of Seattle's voters punched their presidential ballets for Hillary Clinton, overwhelming more conservative parts of the state (and region). Seattle "booms," but "progresses," too. When abstracted from far less convenient facts, its Left Coast reputation is comprehensible.

Like New York, Philadelphia, Chicago, and most big US cities, Seattle has a mayor-council (strong mayor) form of municipal government. Elected citywide without term limits, the mayor heads the city's executive branch, appoints and manages all department and commission heads, and represents the political face of the city locally, nationally, and internationally. Although elections are officially nonpartisan, Seattle's mayors are strongly associated with the Democratic Party (the last "Republican" mayor of Seattle left office in 1969). Similarly, Seattle's officially nonpartisan nine-member city council (seven district-based; two at-large) is at present entirely Democratic in orientation, save the Third District councilor, Kshama Savant, a former software engineer who is a member of the Socialist Alternative Party (and the bane of the Seattle Chamber of Commerce).

Ideologically, Seattle's policy culture is shaped predominately by liberal-democratic ideals and philosophies; moreover, explicitly green and (at times) socialist/social democratic values receive open consideration—limiting business impacts on formal municipal elections (Beekman, 2019). Though Seattle once seemed to fit Clarence Stone's famous model of a "middle-class progressive regime" that places financial demands on business elites despite the structural limitations discussed in Chapter 1, Rosdil (2017, p. 3) has recently argued that "cultural change, and especially the increasing prevalence of moral nontraditionalism within younger age cohorts, has fostered [Seattle's] progressive political instincts"; in this view, "the notion that their future salvation, in the form of climate preservation, depends crucially on what they do today imbues local policymaking with a messianic quality" (p. 12). Rosdil concludes that the negative perceptions of large corporations held by "non-traditional subcultures" in Seattle have created "a vocal constituency for progressive efforts to regulate and extract resources from business investment," even as leading global corporations in Seattle are arguably less anchored to local regimes of urban growth.

Progressive urban policies to address climate change are evaluated in more detail in Chapter 3. They are not, however, the straightforward products of Seattle's culturally progressive voters. "Traditional" forms of corporate pushback are hardly absent, including Amazon's influence on politicians and (as yet, still relatively unsuccessful) efforts to back preferred candidates for city council; redistributive taxes to invest in green urbanism also confront business concerns, even as recent public efforts recognize that "not all businesses are the same, even some of the largest businesses" (see https://crosscut.com/2020/07/what-seattles-new-payroll-tax-says-about-citys-politics). Institutions also matter, both locally and at other scales of authority, as well as across both economic and political domains of power. Seattle's "trait" geographies, such as its relatively low percentage of married households (32 percent) or its exceptionally high percentage of college-educated workers (62 percent), are culturally reproduced through relational geographies of migration, which in turn are shaped by economic path dependencies associated strongly with Seattle's high-tech employment transition. In addition, regional-, state-, and federal-level programs and rules simultaneously enhance and constrain Seattle's urban policies and carbon agenda. In fact, like Freiburg, Germany, Seattle's core sustainability policies partly depend upon how more peripheral areas within the city-region *absorb* the costs of social reproduction, especially in housing markets. Dependent on its relations with others, Seattle's "green" traits cannot be treated like "a discreet geographical container" (Mossner & Miller, 2015, p. 18). Urban self-sufficiency, as Day and Hall (2016, p. 25) note, is a "myth."

Mention has already been made of Seattle's extraterritorial powers in electricity generation and water supply, which span the state. Beyond these formal powers, Seattle's overall development story is impacted by the separately elected Port of Seattle, yet another special-purpose government (like Metro discussed above) that was formed in 1911 and today manages the airport, the main harbor in Elliot Bay, and several marinas.

Though focused mainly on facilitating trade, travel, commerce, and job creation, the Port of Seattle runs multiple environment and sustainability programs, including carbon-reduction initiatives at the Sea–Tac International Airport like rainwater capture, food donations, preconditioned air, and the use of aviation biofuels (Stanton, 2017; Port of Seattle, 2016). Seattle's "progressive" households have recently reduced their reliance on cars faster than all other large US cities (Balk, 2019). Greater density in the urban core has helped. Seattle also benefits from Sound Transit, mentioned earlier in this chapter and discussed more in Chapter 3, as well as regional planning policies associated with the area's federal-designated metropolitan planning organization, the Puget Sound Regional Council (PSRC), that will also be discussed in Chapter 3. An additional issue is how the PSRC's land-use, transit, and technology policies increasingly relate to its emerging climate change and resiliency goals (Puget Sound Regional Council, 2019a).

These formal government institutions and many others (such as the Puget Sound Clean Air Agency, still another special-purpose regional government agency with elected officials) work with a huge array of state authorities, private entities, interest groups, nonprofit organizations, and ad hoc social movements to govern urban carbon issues, choices, and projects in and around Seattle. Representative entities active in Seattle include 350 Seattle, Earth Justice, Got Green!, the Bullitt Center, the Seattle Youth Climate Action Network, Cascadia Climate Action, Climate Solutions, Futurewise, Sightline, and Forterrra (see Box 2.1). These actors have helped Seattle to set carbon agendas, provide access to decision-making arenas, monitor implementation, and, not least, shift the long-term priorities and resources of the region's formal government institutions (Andrews & Edwards, 2004).

Formed in 2016, for example, the "350 Seattle" organization is a 501(c)3 nonprofit associated with the environmentalist Bill McGibbon. 350 Seattle seeks "just transition," defined as accessible family-wage jobs in the transition to clean renewables. 350 Seattle has also advocated for bike infrastructure and lower parking requirements in developments (350 Seattle, 2018). Still other groups, such as Stewardship Partners, have worked with Washington State University and storm water planning groups to build a new regional grid of raingardens, local efforts in the physical transformation of neighborhood landscapes that require what Aaron Clark, their director of strategic partnerships, calls greater attention to intersectionalities between "siloed institutions and local service providers with racial and ethnic communities often left outside of these policies." That goal presumes, of course, that cities with different capacities can break down institutional barriers in climate planning because the nature of climate change, "is not compatible with the specialised and sectorised nature of policy-making institutions" (Oseland, 2019).

Climate Solutions in recent years has focused especially on reforming state transportation funding to incentivize electric vehicles and advocating for a clean fuel standard; it has also worked with other organizations seeking to transition away from fossil fuels. Climate Solutions has furthermore targeted polluted sites and polluted waterways and has an active committee working on shaping Seattle's Green New Deal (Jones, 2020).

> **Box 2.1. Example of Carbon Action Organizations in Seattle**
>
> *Climate Solutions.* Climate Solutions (CS) is a Pacific Northwest region-based clean energy economy nonprofit located in Seattle that seeks to accelerate energy-focused solutions to the climate crisis. CS focuses on "transformational policies and market-based innovations," in partnerships with the Seattle Chamber of Commerce, labor unions, and key government bodies who are all working on new visions of sustainable prosperity. See https://www.climatesolutions.org/.
>
> *Futurewise.* This group was founded in the early 1990s to help support implementation of Washington's Growth Management Act, advocating especially for protecting wildlife habitats, open space, farmland, and working forests by shunting new growth into urbanized areas in ways broadly consistent with Smart Growth regional planning theories and practices. Relevant initiatives include, for instance, the Seattle Climate Atlas & Resilient Voices project.
>
> *Sightline.* An independent, nonprofit think tank in Seattle since 1993 that aims to make the Northwest "a global model of sustainability" by defining sustainability as the intersection of environmental health and social justice. See https://www.sightline.org/.
>
> *Cascadia Climate Action.* Cascadia Climate Action (CCA) in Seattle began in 2014 and now actively maintains "a climate events calendar." They seek to circulate and coordinate public engagement in climate action. Through volunteers, interns, and community collaborations, CCA maintains a website, distributes a newsletter, and sponsors events like the "Climate Science on Tap" series. See http://cascadiaclimateaction.org/.

Seattle's Main Climate Change Challenges and Primary Infrastructural Risks

The challenge of "specialised and sectorised" policy-making—of building up strategic partnerships through new intersectionalities—is one of the fundamental challenges of urban green governance. Benjamin Barber (2013) argued that rescaling global problems like climate change away from dysfunctional nation-states provides new openings for more collaborative, interdependent, transnational urban networks such as C40 Cities and related groups (including the Urban Climate Change Research Network). But American cities are sites of "prior construction," subject to what Orren and Skowronek (2004, p. 96) similarly theorize as "intercurrence," a term that refers to how relatively independent institutions—Seattle City Light, Metro, the PSRC, the Port of Seattle, Sound Transit, the Mayor's Office, 350 Seattle, and the Chamber of Commerce—move "in and out of alignment with one another," *abutting and grating* in political times, over authoritative scales and across social spaces as "multiple orders" in simultaneous action (Dierwechter, 2017; Orren & Skowronek, 1996). What to do given these tensions and challenges?

Traditionally, local comprehensive plans such as Seattle's landmark 1994 *Towards a Sustainable Seattle* confront problems of intercurrence through multi-sectoral visions of publicly deliberated changes that coordinate regulations and investments over a set period of 20 years or longer, albeit through recursive, often annual, updates (Altshuler, 1965). Land use is coordinated with transit, public works support economic development goals, and so on. More recently, smart city infrastructures—for instance, the Internet of Things or interconnected networks of data collection devices—offer synergistic efficiencies through new forms of algorithmic governance and data integration, even as "legal and institutional complexity" remain barriers to the more "interactive management" of city systems and urban spaces (Danaher et al., 2017, p. 17). In reality, alignment that does occur in places occurs through the dogged agency of specific people. Key actors move "from office to office to office to office" getting signatures on key policy recommendations that require multi-institutional agreements and shared protocols (Clark, 2019). In time, these efforts lead to "durable shifts in governing authority" (Orren & Skowronek, 1996) as in Washington, where local comprehensive plans with policy elements like county-designated urban growth boundaries are now mandatory (though "abutting and grating" with older regimes that aim to protect property rights).

The scientific and policy literature on climate change, even when narrowed to the Pacific Northwest and (more narrowly still) the urbanized Puget Sound region, is enormous. Research covers implications for, *inter alia*, systems of water, energy, transportation, agriculture, forests, land cover changes, biochemical cycles, indigenous peoples, and so on. "Urban" sectors in these studies are often separated for purposes of analysis even as "urban" issues are implicated in most "sectoral" dynamics. In consequence, "climate-related disruptions of services in one infrastructure system will almost always result in disruptions in one or more other infrastructure systems" (Cutter et al., 2014).

That said, Seattle's main climate change challenges and primary infrastructural risks relate to *exacerbating* challenges and risks already experienced (Baker, 2020; Georgiadis, 2020). "The most significant changes projected for the Northwest," Seattle's Office of Sustainability and Environment (City of Seattle, 2017a) reported in 2017, "include changes in temperature, precipitation, sea level, and ocean acidification. Flooding, heat waves, and extreme high tides are not new challenges in Seattle. [...] However, climate change will shift the frequency, intensity, magnitude, and timing of these events" (p. 29).

"Water," in a word, is the transcendent *natural* theme of local concern. The host of Seattle's specific risks include: (1) greater variability in precipitation (wetter falls, winters, and springs; drier summers) and more extreme precipitation events (with the top 1 percent 24-hour rain events expected to be over 22 percent more intense, on average, by the 2080s); (2) a projected sea-level rise of two feet, on average, by 2100, with the normalization of attendant storm surges; (3) increased frequency and scale of landslides due to the city's post-glacial geology, steep slope topography, and wet(ter) winters; (4) pronounced difficulties with urban drainage systems, including sewerage overflows; (4) erosion, salt water intrusion, corrosion, and loss of near-shore habitats; (5) increased risks of vector-borne, foodborne, and indeed waterborne diseases, with disproportionate impacts on people with existing medical conditions, people of color, low-income communities, immigrants, refugees, and homeless populations (pp. 11–12).

"Injustice," then, is the transcendent *social* theme of concern. Seattle's overarching climate efforts to reduce emissions and prepare for impacts accordingly prioritize actions "that reduce risk and enhance resilience in frontline communities (e.g., communities of color, lower income communities, immigrant and refugee communities, disabled residents and seniors)," whose climate vulnerabilities result from "decades of systemic exclusion from power" (City of Seattle, 2017a). Social equity concerns are not new; however, a comprehensive review of policy documents suggests they are more prominent than in past plans and policy statements.

For example, while equity issues are inferred in Seattle's 2006 Climate Action Plan, they are repeatedly invoked in the more recent 2013 Climate Action Plan, even as this plan—like most plans (Ford, 2010)—works with and through a policy "ecosystem" of other municipal plans, including transportation and land-use plans, building and energy plans, waste plans, as well as "climate-change related" plans on disaster readiness and hazard mitigation (p. 6). Accordingly, race and social equity is the "foundational core value" of Seattle 2035, the emerging comprehensive plan for the city going forward (City of Seattle, 2016b). This will constitute an especially important theme in the next two chapters.

Conclusion

Seattle's green development story is unfinished. The select history recounted here is indicative but not determinate, which provides at least some hope for the still-open possibilities of a post-carbon future that is at once ecologically resilient and socially just. It will not be easy. Stimulated at different times in its relatively short history by gold, airplanes, warfare, software, and of late the e-digital prowess of Amazon, Seattle has grown and changed through cycles of boom and bust, even as it has retained a discernable tradition of political and policy reformism often informed by radical critiques and social movements.

Tensions that pervade these traditions continue to reverberate and will do so in the future. Billionaires have recently remade the urban core of Seattle into a high-tech space for what Allen Scott calls "the cognitariat," a process of reurbanization that has helped to alter the race and class structure of the city— and, I insist, the wider city-region as a whole (Dierwechter, 2017; Herrschel & Dierwechter, 2018). This has led to multiple contradictions and challenges (Gibson, 2004). Seattle is highly educated, politically progressive, even here and there radical, as the post–George Floyd promulgation of the "Capitol Hill Occupation Protest" (CHOP) in June 2020 revealed. Seattle attracts but also questions corporate power. Seattle has embraced a huge range of urban sustainability initiatives, reduced its municipal reliance on cars, and drawn the favorable attention of researchers for decades. Yet its homeless camps mushroom, its black population collapses, its suburban frontiers choke in the fumes of congestion. Sustaining Seattle has meant reproducing geographies of pleasure and pain, of profit and pollution (Gibson, 2004; Dierwechter, 2017).

Negotiating these long-running contradictions, therefore, will be at the center of building a carbon-friendly city in what remains an unsustainable world. While both the 2006 and 2013 plans focused on mitigation efforts, the 2013 plan deepened local policy concerns with the urgency of adaptation measures within the political context of social equity goals. These measures help communities and ecosystems to cope with changing climate conditions. By 2017, adaptation measures permeated urban climate policies (City of Seattle, 2017b), notably in major service areas like regional drainage systems and water supply. Concrete examples of adaptation have included developing new strategies for innovative approaches to managing stormwater, conducting various threshold analyses of sewer networks to determine sensitivities to new storm events, and assessing vulnerabilities of forested watersheds (p. 26). In 2016, Seattle City Light's integrated resource plan had identified future investments in renewable energy as "the most cost-effective way to meet demand and [even] reduce reliance on hydropower" (p. 56). By 2020, Seattle was clearly searching in different ways for Klingle's new ethic of place, "one grounded in history" (ibid.) but with an eye on a precarious environmental and social future.

Chapter 3

CURRENT SITUATION: BUILDING A "CLIMATE-FRIENDLY" CITY IN AN UNSUSTAINABLE WORLD

> Climate change is not a stand-alone issue separate from the other issues Seattle faces. It is rooted in land use, transportation, and building energy patterns that have evolved over generations, and therefore, the solutions to climate change also cannot be stand-alone. They must be part of Seattle's work to build vibrant, complete communities, and they will require action from everyone in our community—local government, residents, businesses, industry, building owners, utilities, and many others—as well as action at the state, federal, and international level.
>
> City of Seattle (2013a)

Seattle's political history has been characterized by climate change discourses and related urban sustainability concerns for many years now (City of Seattle, 1994, 2009; Lee & Painter, 2015). Since the early 1990s, Seattle's civic leaders have explicitly considered climate mitigation strategies (Bassett & Shandas, 2010; Brunner, 1991). The municipal government today employs public officials who focus on how global climate change is affecting (and will affect) the resiliency of infrastructure systems, the built environment, urban form, transportation choices, and many other critical policy arenas. In addition, the city of Seattle participates in local, national, and global climate action networks. These include the King County–Cities Climate Collaboration (K4C), the Mayors Climate Protection Agreement (MCPA), C40 Cities, the Global Covenant of Mayors, and Carbon Neutral Cities Alliance. Trans-local networks such as these help to shape a multitude of local relationships with nonprofits, key foundations, and industry associations as well as state and federal authorities.

Based on the analysis of municipal archives, online published documents, and public studies—notably the city's climate action and ancillary plans

(City of Seattle, 2006b, 2013b, 2016a, 2017)—along with a range of local and nonlocal newspaper accounts over the past 30 years, the discussion here explores urban climate action as a challenge important to the management of Seattle in the spatial context of the wider city-region.

Challenges are socially constructed, as Michael Callon (1984) initially argued, through "problematization." This is the first step in building what he called actor networks for the eventual "mobilization" of new realities, including the transformational project of post-carbon cities that can meet the demands of global ecological change. As Rutland and Aylett (2008) note in their important study of how climate action policies ultimately surfaced—or "translated"—in Portland:

> When individuals come to view themselves and their goals according to the same metrics as the [local] state, and base their actions on these metrics, they become part of [a] network of self-regulating actors. [...] This "translation" of interests [...] binds people together through a shared perception of reality and through the codes of conduct that seem to flow naturally from it. (p. 630)

That may not occur. Change is elusive. "Translation" is constant, precarious, reversable. New orders—new kinds of cities, for instance—are better understood as the never-ending micro-capillary work of *ordering* complex, heterogenous "sociotechnical systems," as the ARC3.2 framework discussed in Chapter 1 puts it. Socio-technical systems fall apart, decay, splinter, and/or simply refuse to congeal into spaces mobilized around consolidating new realities. Still, change does occur. Focusing on Seattle's efforts to become a more climate-friendly city in an increasingly unsustainable world, I argue below that important developments in carbon mitigation and adaptation efforts nonetheless exemplify mounting political concerns with social polarization and racial inequities and thus noteworthy dissatisfaction with past and current climate policy progress.

Green Shoots: From "Environmental Conservation" to "Urban Sustainability"

Like many cities and regions in the United States, apprehensions with environmental conservation have a long history in the Seattle area. Cleaning up Lake Washington was the rallying cry in the late 1950s. The politics of environmental conservation—what Samuel Hays (1959) famously called "the gospel of efficiency"—also marked Seattle's public reforms in the Progressive Era (1890–1920). For example, in 1909 the Washington Conservation Association hosted the first National Conservation Congress as part of the Alaska-Yukon World's Fair. Attendees discussed classic topics like irrigation, soils, good roads, mining, and forestry. They also broached problems in political economy like "the relation of Capital to Labor in the work of general conservation of natural resources" (Ott, 2008).

The journey from a traditional, largely reactive, environmental conservation discourse to a more proactive urban sustainability agenda, much less explicit climate action, reflected these original tensions in subsequent decades. In a special congressional election held in 1977, a Republican candidate, Jack Cunningham, upset his Democratic rival, Marvin Durning, by painting him as an "environmental extremist." Portending a now familiar pattern across the United States, Cunningham won the "Democratic blue-collar precincts of southern Seattle" (Cannon, 1977, p. A2). As discussed in Chapter 2, Seattle was still a shrinking city in 1977. Protecting blue-collar jobs triumphed easily over promoting environmental concerns, at least for national office. Yet the "roots of urban sustainability" would soon burst through the political landscape during the 1980s, intertwining with economic, racial, and gender equity concerns to grow into what Jeffrey Sanders (2010) has called the city's "grassroots counterculture"—or what I have elsewhere theorized as Seattle's "radical-societal political order" (Dierwechter, 2017).

This radical, more activist, order crystallized in the 1960s and early 1970s. Anti-highway protesters, for instance, gained considerable some ground on the region's pro-growth coalitions (Steven, 2011). More recently, Seattle garnered global attention in 2015 as groups of colorfully-clad kayakers gathered in Elliott Bay to protest Artic drilling plans by Shell Oil; dubbed "kayaktivists," Burkett (2016) shows how this novel form of "climate disobedience" has moved urban green activism "beyond the courtroom and legislative hallways" (p. 3). "Radicals" have never actually governed Seattle, much less the metropolitan region. Still, Seattle's "grassroots counterculture" has (selectively) pressured "state-progressive" institutions of regulation and investment otherwise pulled to the right by market-oriented interests and the accumulation order (Dierwechter, 2017). In his work on the urban politics of mobility, Kevin Ramsey (2009) has shown that Seattle's activist groups during the early 2000s were successful in establishing the climate crisis as an issue of general concern in wider debates about major infrastructure redevelopment on the city's waterfront, notably the Alaska Way Viaduct, but that, in the end, they were mostly unable to redirect established regimes that tended to perpetuate automobility.

What we see, then, are more like "green shoots"—some impressive, others less verdant—rather than a comprehensive "green canopy" enveloping a new kind of millennium city, especially when "Seattle" is extended into the more unsustainable city-regional space-economy and beyond (Figure 1.2). That duly acknowledged, by the mid-2000s the city of Seattle was habitually celebrated by nonlocal elected officials and academics alike for its political leadership on urban sustainability and, in time, global climate change as well (Cornwall, 2007; Portney, 2003). Why? How did Seattle come to occupy this reputational position among observers of urban sustainability and, later on, progressive climate action?

The "Carbon Turn": From Urban Sustainability to Global Climate Action

Urban sustainability in Seattle is less an "add-on" domain of territorial governance and city–society relationships than an interrelated set of goals and processes that (try to) permeate *extant* civic and private institutions with their own internal path dependencies, values, norms, and modes of operation. It is both ubiquitous and hard to pin down.

Offices of Sustainability in American cities are often "flags without countries," working to promote innovation adoption by building consensus across otherwise siloed entities (Laurian et al., 2017). Urban sustainability, in practice, therefore seeks, *inter alia*, new forms of economic development; novel ways to provide parks, open space, and youth education; improved energy provision, public purchasing, and civic investments; better linkages with regional food and agriculture complexes; forestry protection and street greening; cleaner transportation fleets with denser routes; the deployment of green buildings and different development codes; more efficient waste management, water and wastewater systems; and, not least, smarter land-use planning systems and urban forms. It was (and is) everywhere, but also for skeptics nowhere in the deeply transformative sense—more like "rain without thunder," to borrow from Frederick Douglass's elegiac critique of American progress.

In recent years, climate change efforts arguably have displaced urban sustainability goals as the organizing governance leitmotif for greener urban futures. Goals and processes remain similar for these core policy arenas of urban development and societal change. Yet getting to urban sustainability as a hypothetically balanced form of development—and specifically imagining its everyday practices in space and society—was an important core "framing" for the climate focus of recent decades, in Seattle as much as anywhere. As Figure 3.1 suggests, adaption to climate change "works through" broader, more abstract sustainability values that seek to integrate ecological, social, and economic aspirations. The key policy moment in this conceptual and practical relationship in Seattle is arguably the most important plan in Seattle's urban planning history—the municipality's landmark 1994 comprehensive plan (City of Seattle, 1994).

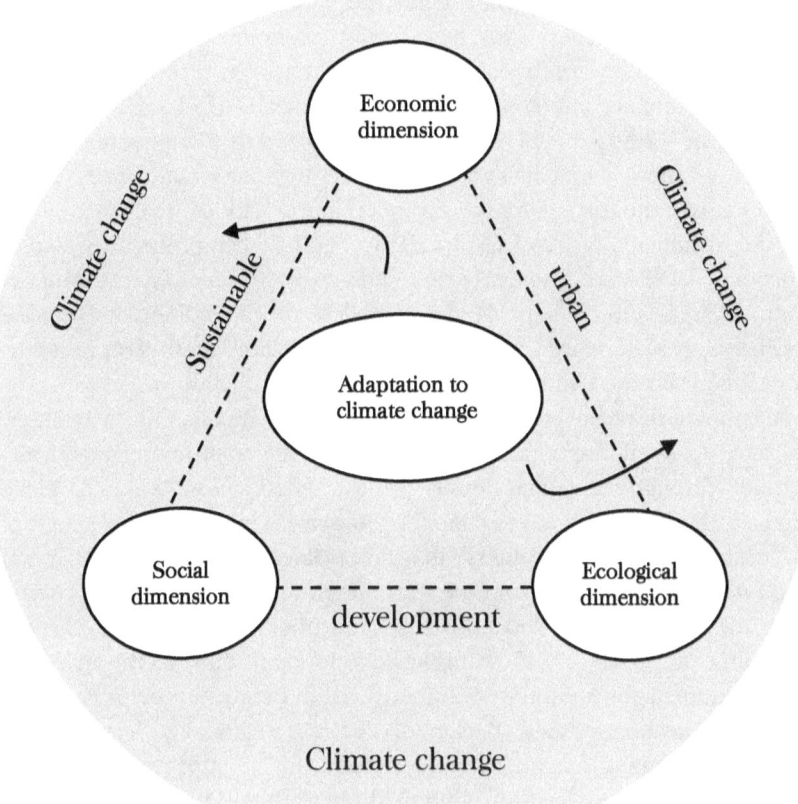

Figure 3.1. Sustainable urban development and climate change (Source: Interpreted from Yohe et al., 2007, p. 815)

Toward a **Sustainable Seattle:** *A plan for managing growth, 1994–2014*

A complex, integrated, multidimensional 20-year policy plan as much as a physical rendering of future land-use patterns and spatial forms, *Sustainable Seattle* reflects the legal impress of top-down state legislation as well as, in theory, bottom-up citizen participation. Also important are the conceptual powers of professional theories, such as Smart Growth and New Urbanism, to shape more grounded practices (Whittemore, 2015).

Mandated by the state's 1990/91 Growth Management Act, this comprehensive plan for land use, housing, transportation, and other key policy elements focused squarely on where and how to accommodate renewed growth forces associated with post-Fordist economic restructuring in the 1980s (Dierwechter, 2008). *Sustainable Seattle* has been repeatedly amended, with formal updates in 2004/5 and, most recently, 2015/16 (City of Seattle, 2016b). The 1909 plan for Chicago—the Burnham Plan—is remembered for its civic embrace of Chicago's regional setting; *Sustainable Seattle* is historic because it was the first comprehensive plan in the United States to aim explicitly for sustainability (Abel et al., 2015). Seattle's comprehensive planning efforts since 1994 are frequently touted as "best practice" by the American Planning Association. The 1994 plan figures in the fifth edition of *Urban Land Use Planning*, the standard training text for urban planning students in master's programs across the United States (Godschalk et al., 2006).

In contrast to recent strategies for sustainability in places like Minneapolis that have upzoned more broadly across urban space to promote greater racial inclusion through residential density (Inouye, 2020), *Sustainable Seattle* elected originally—and still elects—to channel new growth into a targeted geography of "urban villages" and "centers" that mix residential and commercial activities. Locally, this strategy from the start aimed to protect roughly 70 percent of Seattle's (in the main) "open low-rise" morphology (Stuart and Oke, cited in Raven et al., 2018, p. 145) from new development through the legal shield of conventional single-family zoning, a decision bemoaned by many climate and social justice activists (Beekman, 2019a). Regionally, Seattle's urban villages have sought to accelerate wider visions for an improved public transportation network, especially around light rail stations (Puget Sound Regional Council, 1998).

The urban village model anticipated and then neatly exemplified key tenants of the "smart growth" movement, the planning profession's perennial quest for "balanced" development given diverse economic interests and competing cultural demands, the not in my backyard (NIMBY) power of Seattle's homeowners, but in addition, and not least, a mounting sense that cities through coordinated, multi-scalar, planning interventions had to step up to the global challenge of sustainable development (Cowell & Owens, 2010).

The most recent version of *Sustainable Seattle*—*Seattle 2035* (City of Seattle, 2016b)—remains committed to this model. The plan suggests the "volume" of the city's current regulatory footprints can accommodate new growth for at least several more decades. But *Seattle 2035* is more concerned with equity and justice problems now destabilizing the city and region, particularly the related issues of racialized displacement and housing unaffordability (Beekman, 2019a; Fowler, 2016). This is captured by the synoptic goal of managing growth to become an "equitable" *and* sustainable city (ibid.). As amended, *Seattle 2035* seeks to encourage alternative affordable home ownership through, *inter alia*, community land trusts, down payment assistance, mixed-income housing requirements, and limited-equity housing co-ops in addition to more traditional developer-oriented tools like density bonuses, increased floor area ratios, and transfers of development rights from King County's peripheral areas slated for ecological conservation (Rush, 2016; McCormick, 2017). Ordinance amendments in 2019 to *Seattle 2035* further increase development capacity in the city's 27 urban villages while explicitly "hard-wiring" requirements for more affordable housing by allowing developers and builders to either build more densely on-site or to pay into an affordable housing trust fund.

The growing equity and social justice concerns in formal planning discourses over zoning codes, transport investments, and housing programs have further merged with the city's focus on climate policies. Planning for climate change and alternative forms of liveability have both gained the serious attention of local activists, the business community, elected officials, and Seattle's professional *cognitariat* both inside the municipality and across other public, private, and nonprofit institutions operating at various scales of authority and agency (Hinshaw, 2016).

As discussed briefly in previous chapters, local discussions of global climate change were evident by the early-1990s. *Sustainable Seattle*, for instance, had linked urban villages with carbon reduction. Such villages are, in theory, more self-sufficient neighborhoods that coordinate housing and employment opportunities near public transportation options. But the evolution of local planning for climate change accelerated in the aftermath of the COP-3 negotiated Kyoto protocols, ironically as much for their federal rejection by George W. Bush as their local embrace by Seattle's civic leadership. In the politically and ecologically disastrous Trump era, these efforts intensified, as discussions of a "local green New Deal" strongly suggest (City of Seattle, 2019; Tigue, 2019; Jones, 2020). Still, the constant work of enrolling and mobilizing diverse agents into a common space for local carbon efforts illustrates the political challenges of urban transformation first broached in Chapter 1.

The Twenty-First Century Evolution of Planning Justly for Climate Change

The empirical shift to climate action, whether concerned with mitigating carbon emissions from cities or adapting cities to global climate change, is not easily dated. No "Pearl Harbor" moment stands out. Global accords to reduce greenhouse gases, especially the Kyoto Agreement signed in 1997 and formally ended in 2012, provide a popular line of investigation, as countries have mapped out various national strategies. However, the United States officially dropped out of the Kyoto Agreement in late 2001. Instead of a coordinated national policy on climate change, as seen in Germany, for instance, an uneven and poorly understood patchwork of state and local initiatives tried to fill the gap (Bassett & Shandas, 2010; Deetjen et al., 2018; Dierwechter, 2010; Osofsky, 2015).

Generalizations are difficult. Various states, such as Florida, initially strengthened statutes that required energy conservation and emissions reductions but have since dramatically reversed course. In contrast, California mandated conservation elements in local plans and has since strengthened them. At the municipal level, by the mid-2000s, explicit climate action plans (CAPs) in larger US cities like Denver, Portland, and Seattle started to "break away" from the environmental and/or conservation elements within local comprehensive plans as shaped by enabling legislation to become stand-alone statements of urban policy and carbon management in their own right (Hughes, 2016; Mason & Fragkias, 2018).

Seattle's two major CAPs are analyzed here: the 2006 CAP, which emphasized meeting the Kyoto goals through carbon mitigation strategies, and the 2013 CAP, which thus far has focused even more on adapting to the reality of climate change through enhanced resiliency. The discussion specifically considers key institutional changes within Seattle that this carbon turn has abetted. The larger context within which this shift has occurred is crucial to understand. Following Hess (2019), Seattle's prominent carbon turn in urban politics and policies represents an American exemplar of the "challenger city," working in effective opposition to inconsistent and often hostile federal policy. Despite the welcome hiatus of the Obama years, this climate agenda was less influenced by professionally and scientifically coordinated multilevel governance than the intercurrence of local institutions and the mounting power of trans-local municipal networks.

The 2006 climate action plan: Meeting Kyoto's challenge through carbon mitigation

Spearheaded by a citizen-based green ribbon task force, Seattle's first ever CAP in 2006—*Seattle, a Climate of Change: Meeting the Kyoto Challenge*—was a manifesto of 18 recommendations "to reduce global warming pollution across the community" (p. ii).

This departure had continuities. Reducing pollution was hardly novel. And many of the recommendations in the plan were already moving forward in various public institutions. Seattle had long been trying to "reduce its dependence on cars" (p. 5), "increase fuel efficiency and use of biofuels" (p. 10), and "achieve more efficient and cleaner energy for […] homes and businesses"

(p. 16). Specific recommendations, such as expanding "compact, green, urban neighborhoods" (p. 9), amplified strategies like the "urban village" growth concept just discussed. Yet these otherwise familiar arenas of governance were nonetheless reworked into a kind of actionable theory for how the city might best combat global warming in the near term—and especially how the city might meet Kyoto goals through local carbon mitigation. Commercial parking taxes; a road pricing scheme; bicycling infrastructure; improved multi-scalar and regional partnerships; reducing emissions from diesel trucks, trains, and ships—all of this suggested the fast-growing policy importance of new carbon-reducing relationships with the economy, the materiality of urban space, institutional routines, and daily flows within, through, and well beyond the city.

Meeting the Kyoto Challenge is an example of what Robert Beauregard (2012) calls "planning with things." Things—"apartment buildings, site plans, scale models, and parking spaces"—problematize to generate interest and, then, if/when agents can be successfully enrolled, help to mobilize new realities. "For planners to act with authority and certainty," he concludes, "they must create alliances, and to create alliances they *must gather people and objects to them.* By enrolling people and things in their projects [and plans], planners become influential" (p. 188), where "planners" also include citizen planners. *Meeting the Kyoto Challenge* is furthermore an example of what Jennifer Rice (2010, pp. 932–33) soon thereafter called "the climatization of Seattle's urban environment" and the "carbonization of urban policy." Both developments helped the city of Seattle to plan with and territorialize that ultimate "thing" of all—the greenhouse gas (GHG) molecule, which was now made a normal "object" of governance:

> With the use of GHG inventories and the centralization of local environmental policies around issues of climate change, Seattle has constructed political authority with relationship to the climate through the territorialization of carbon. These state practices work to embed the abstract concept of climate change onto the urban environment and into city politics. [...] Although it might seem that climate represents an unterritorializable form of material because once emitted, emissions become part of the globally mixed atmosphere, the city of Seattle has created new strategies for incorporating greenhouse gases into its territory. (p. 934)

A 2016 report on Seattle's community-wide GHG emissions in the transportation, buildings, waste, and industrial sectors suggested that the interrelated processes of climatization, carbonization, and territorialization had helped to reduce emissions by 5 percent while the city's population grew 18 percent.

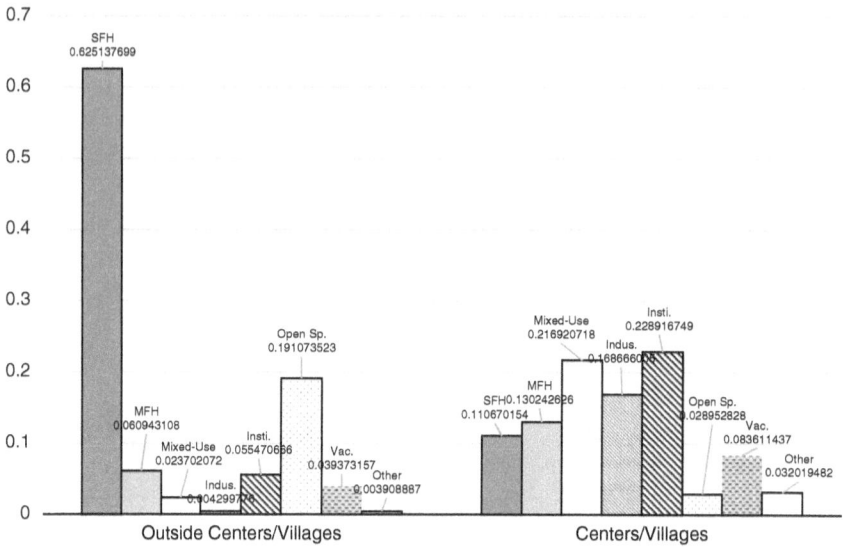

Figure 3.2. Territorializing carbon action: Net land use acreage in Seattle in 2016, both "inside" and "outside" the city's urban centers and villages (Source: Data calculated from Land Use Appendix Table (City of Seattle, [1994, 2005] 2015)

This resulted in a decrease of 20 percent in per-person emissions. With an annual per-person emission rate of 4.3 metric tons of CO2e (carbon dioxide equivalent gases), the report concluded, Seattle had become "one of the most climate friendly cities in the nation" (City of Seattle, 2016c). Seattle's CO2e figure for 2016 was comparable to Sweden's 4.19 CO2e rate in 2017 and much lower than Iceland's rate, a country with half the population of Seattle (Richie & Roser, 2017).

In practical terms, climatization, carbonization, and territorialization are manifested in changing local land-use configurations (Figure 3.2). Seattle's long-targeted villages and centers deemphasize single-family housing (SFH), valorize commercial mixed-use spaces (Mixed-use), and, albeit only in select "centres," preserve more industrial uses (Indus.) than areas outside "villages and centres"—an important element in "keeping blue collars in green cities" (Dierwechter & Pendras, 2020; Holmes, 2020). So, while Seattle as a whole is still defined mainly by the detached, garage-enhanced, single-family home of twentieth-century American urbanization and urban culture—or "open low rise" design (Stuart and Oke, cited in Raven et al., 2018, p. 145)—climatization and carbonization have slowly led to the material emergence of new territorial forms, albeit more incremental than transformative in pace.

Certainly, many other policy developments, including green energy programs, regional transit lines, and water management systems, also contribute decisively to CO2e reductions. But Seattle's centers and villages are arguably still most identified with the city's climate policies and urban sustainability efforts (Godschalk et al., 2006). Building energy use, for example, represents one-third of Seattle's GHG emissions, and since 2005 the city has benefitted from Seattle City Light's "carbon neutral electricity" (City of Seattle, 2018, p. 16). Yet much as residential suburbs rely on primary cities to mitigate carbon emissions (see Dierwechter, 2010), primary cities rely on select areas for their carbon action (Barbour & Deakin, 2012; Dierwechter & Wessells, 2013; Osofsky, 2015).

At the same time, Seattle's much-touted spatial planning strategy remains difficult to assess empirically. One recent mayoral statement, for instance, has claimed that "smart growth is the foundation of effective transportation policy, and Seattle's nationally recognized urban village strategy, adopted in 1994, provides the essential foundation for Seattle's climate-friendly transportation policies" (City of Seattle, 2018, p. 8). Yet the same statement also frankly admitted, "[Seattle] does not current[ly] have a city-wide policy for tracking the carbon impacts of policies or projects, which limits our ability to fully understand the cost or savings related to carbon emissions" (p. 16). Approximations from general studies elsewhere provide some helpful insights. According to a New York Metropolitan Transit Authority study in 2009, for instance, single-family homes in urban areas of the United States consume 66 percent more energy on average per year than multi-family homes as measured in millions of British thermal units (*m*BTUs). Single-family homes in *suburban* areas outside cities consumed nearly 200 percent more *m*BTUs per year than multi-family homes (Blue Ribbon Commision on Sustainability, 2009).

The 2013 climate action plan: Adapting with just resiliency

Waiting for sufficient data to pile up before acting, however, is not possible. The price of political inertia is now all but unfathomable, with human self-extinction as one entirely plausible future (Claire, 2013). Yet several caveats are important to highlight. The otherwise impressive rollout of carbon-reduction policies in cities like Seattle generates its own set of unevenly experienced costs, notwithstanding the likely savings associated overall with lowering CO2e emissions—known and unknown.

For all its progress, by the early 2010s, urban sustainability and climate action policies in Seattle, as elsewhere in the United States, were struggling to produce socially inclusive spaces of ecological protection. Academic studies of Seattle over the past decade have habitually mapped the city's new "carbon territories" (Rice, 2010) with terms like "ecological gentrification" (Dooling, 2009), "smart segregation" (Dierwechter, 2014), "eco-gentrification" (Rice et al., 2019), and "gentrified sustainability" (Abel & White, 2015). Gentrification and worsening segregation are exceptionally high costs to pay for green dividends.

This is hidden in plain view. Urban practitioners and activists across Seattle, inside and outside the state, are acutely aware of Seattle's growing reputation as an "elite emerald" rather than "emerald city" (Dierwechter, 2017, p. 35). As one article has noted of the city's public schools:

> Seattle public schools are becoming more divided when it comes to race. For years, the city made an effort to integrate schools with a busing program, to mixed success. But while residential desegregation continues to progress in Seattle, schools have been resegregating over the past few decades. Because of this, a school's student body can look very different depending on what neighborhood it is in. (Dev & Brazile, 2019)

Maps of risk displacement by neighborhood further indicate the scale of this concern (Figure 3.3). Recent nonprofit workshops on building green infrastructure, for example, have included concerns with intersectionality, exploring how social inequalities of race, class, gender, sexuality, age, ability, and ethnicity are refracted through rain gardens, bioswales, water systems, and low-impact development regulations (Stewardship Partners, 2020). The City of Seattle (2016a) has developed an "Equitable Development Implementation Plan" as a detailed and focused complement to traditional planning and development policies (Bakkenta, 2020). The Puget Sound Regional Council (PSRC), the region's Metrpolitan Planning Organization, and Sound Transit, Seattle's regional transit agency, have focused on equitable transit-oriented development (TOD) strategies, in part as a response to state legislation requirements enacted in 2015 (Sound Transit, 2018). Projects include partnerships with nonprofits in the historically diverse Rainier Valley. Puget Sound Sage, for instance, has worked with Sound Transit in recent years to alleviate the displacement effects of traditional TOD, arguing that "in-movers own cars at high rates" while "low-earning residents use transit more frequently to get to work" (Puget Sound Sage, 2012, p. iv).

Inclusive spaces of change are not simply moral achievements. They are ecological imperatives (Raven et al., 2018, p. 165). Low-income citizens in cities like Seattle are often the best agents of carbon mitigation, even as "working class sustainability" projects and environmental justice movements do not receive the environmental credit they deserve (Bell, 2019).

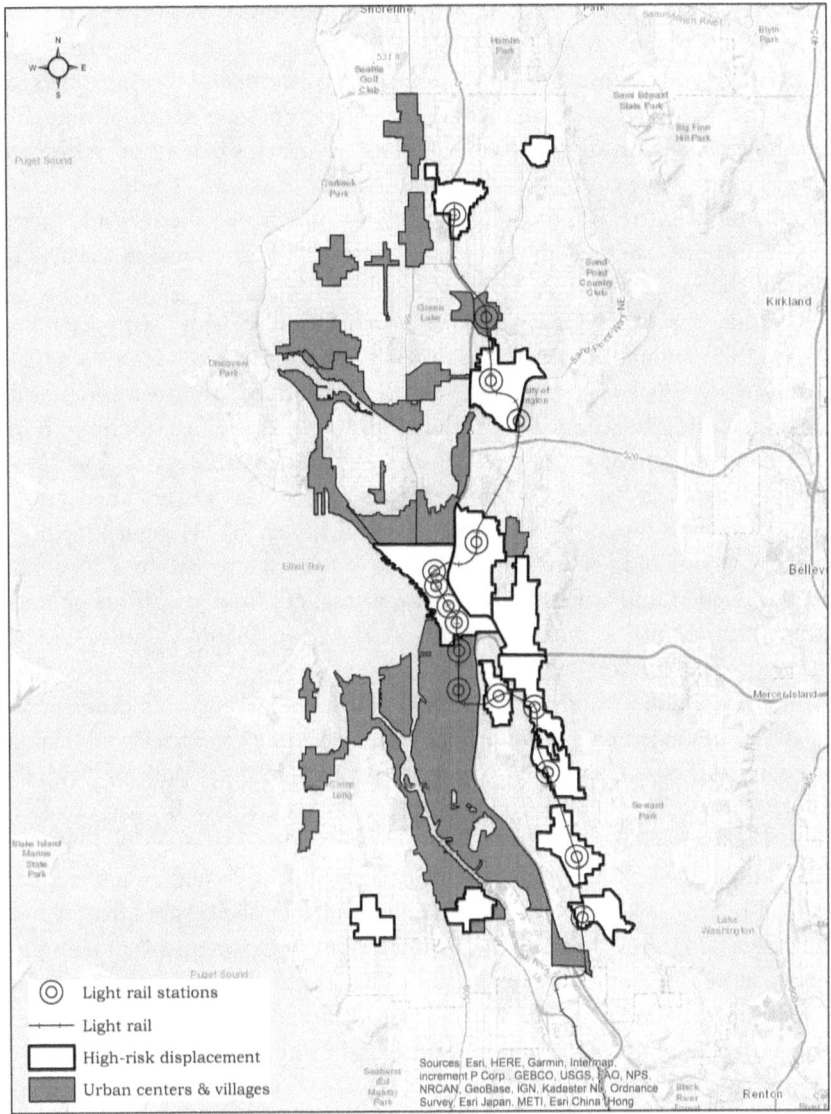

Figure 3.3. Areas of Seattle at "high risk" of social displacement (Source: Author's rendering)

Social justice and racial equity themes permeate Seattle's 2013 climate action plan (Pacino, 2014; City of Seattle, 2013a). The plan seeks to build projects with, as one example, Environmental Professionals of Color (EPOC). The main goal of EPOC is to strengthen the environmental movement by cultivating leaders of color who are working on conservation, climate change, sustainability, and environmental health. Another group includes the Raging Grannies, who, much like Code Pink or the Grey Panthers, use humor, music, and activism to raise local awareness of climate change.

Overall, the 2013 CAP seeks to provide Seattle with "a coordinated strategy for action that cuts across City functions, and focuses on City actions that reduce GHG emissions while also supporting other community goals, including building vibrant neighborhoods, fostering economic prosperity, and enhancing social equity" (City of Seattle, 2013a, p. 3). The plan's opening question and thematic refrain—"How can we enhance equity through climate action?"—focuses on the triptych of (1) transportation/land use, (2) building energy, and (3) waste management, "where City action is most needed and will have the greatest impact" (ibid.). But this opening question might just as easily be reversed: How can Seattle enhance climate action through equity?

The plan's policy framework for transportation/land use, building energy, and waste management continue to emphasize Seattle's *mitigation* goals in light of equity and justice concerns. Illustrative transit/land-use policies "with the greatest impact" include ongoing efforts to improve sidewalks and crossings on arterial streets in order to connect the city's urban centers and villages, while adding new "smart city" electronic real-time bus schedule information and off-board payment options. Building energy policies similarly include smartness programs, such as the technological deployment of more smart meters in homes that can more efficiently provide real-time energy use information to Seattle City Light, who are also charged in the plan with advancing a program to provide both financial and technical assistance to tune up energy systems in existing commercial buildings. Finally, the city's perceived waste policies "with the greatest impact" on carbon mitigation goals range from the banal (e.g., opt-out programs for junk mail) to the more systemic (e.g., phase-in bans on waste from job sites and transfer station, such as recyclable metal, cardboard, plastic film, carpet, clean gypsum, clean wood, and asphalt shingles).

What stands out in the 2013 CAP, though, is the rising challenge of resiliency. In particular, the plan starts to emphasize more explicitly the urgency of urban adaptation to the reality of global climate change, with the aim to develop a comprehensive adaptation strategy. Seattle's lead climate agencies—Seattle City Light; the Office of Sustainability and Environment (OSE); the Office of Planning and Development; the Office of Neighborhoods, Public Utilities, Parks, and Recreation; Public Health; and the Department of Transportation—are now charged formally with "preparing for climate change" (City of Seattle, 2013a, pp. 13–14). Hard climate science projections, dealt with in more detail in Chapter 4, are mobilized to rationalize the plan's policy imperatives: by 2040, the plan reports, average annual temperatures in Seattle will increase from 1.5°F to 5.2°F; by 2080, from 2.8°F to 9.7°F. Puget Sound's mild summers today could be as much as 12.4°F warmer on average by 2080—much more like Los Angeles in Southern California than the Pacific Northwest, with wetter winters and drier summers.

The language is more ominous, more sober, if still laden positively with concerns over social justice and racial equity that were discussed above. For the OSE, for example, the city should

> conduct a citywide assessment of the impacts of temperature, precipitation, and sea level rise on City infrastructure, operations, facilities, and services, including human health with special attention to vulnerable communities. (p. 13)

For the Office of Planning and Development, as a second example, the city must

> evaluate the impacts of sea level rise on flood prone areas and shoreline development and habitat, and consider implications for land use management strategies. (p. 14)

Future sea-level rise is inevitable, the main warrant for current action. The water will find its way into the most exposed crevasses of the city. Fluctuations in temperature and precipitation will grow. The impacts of climate change on public health will deepen, with "disproportionate impacts on lower income, recent immigrant, older, and very young residents, who are at greater risk of health impacts from climate change" (p. 14). In consequence, the organizing question in the 2013 CAP has fundamentally changed—from "How can we enhance equity through climate action?" to "Why and what do we need to prepare for a changing climate?" (p. 54).

Preparing for a Changing Climate: Institutional Responses

The steady greening of urban policies and plans, the environmentalization of the median citizen's expectations of elected officials, and the material production of carbon territories—all of these suggest important empirical changes in Seattle. But to what extent, and how, have recent public preparations for global climate warming actively reshaped the political development of the local state? In addition, how have these preparations intersected with civil society and the regional economy?

In the early 1990s, scholars started to argue that "environmentalism," however defined or mobilized, was joining "conservatism" and "liberalism" as a central political discourse in advanced economies (Peattie & Hall, 1994). That now appears overly sanguine as a general statement. Other observers have noted more accurately that political developments in the variegated state, particularly across the federated United States, involve a diverse array of unevenly experienced changes (Sellers, 2002). These include the impacts of policy entrepreneurship and mayoral leadership, institutional capacity, management cultures, and, not least, persistent organizational silos (Bulkeley & Betsill, 2005; Laurian et al., 2017). Still, local institutional responses matter. They have clearly evolved in major American cities like Seattle in recent decades. This is crucial to understand here. As Hanson and Lake put it,

> Sustainability is fundamentally a political rather than technological or design problem. […] The greatest barrier to sustainability lies in the absence of institutional designs for defining and implementing sustainable practices in local contexts. (cited in Laurian et al., 2017, p. 271)

Consider leadership first. Benjamin Barber's (2013) provocative thesis that big city mayors should "rule the world" and to some extent "already do" is overstated. Nonetheless, it captures the rise of green mayors, especially those with strong executive powers, for example, Michael Bloomberg (New York), Iñaki Azkuna (Bilbao), Bill Peduto (Pittsburgh), Ulrich Maly (Nuremberg), and Anne Hidalgo (Paris), as important political leaders and "policy entrepreneurs" in global climate debates in recent years. In fact, defining and implementing sustainable practices and carbon initiatives in local contexts depends in no small measure on supportive mayoral leadership, especially when linked to more progressive voting coalitions that put "activists in city hall" (Clavel, 2010). "Look at any [active] city in the world," Krueger and Agyeman (2005, p. 415) suggest in their early review of what they called the "actually-existing sustainabilities" of urban capitalism, "and there is a courageous Mayor."

Significant evidence exists for the steady "greening" of the mayoral office in Seattle (City of Seattle, 1997; Office of Sustainability and Environment, 2002–2005). Norm Rice (1990–98), Seattle's first African American mayor, oversaw the 1995 comprehensive plan that championed urban sustainability. Greg Nickels (2002–10), discussed more below, cocreated the MCPA in 2005; Nickels also established an Urban Sustainability Advisory Panel in 2002, which led to the city's first CAP in 2006, an effort that drew on a Green Buildings Task Force. Ed Murray (2014–17), Seattle's first openly LGBTQ mayor, oversaw and supported Seattle's second major CAP. Finally, Jenny Durkan, who followed Murray in 2017, continued to champion climate policies (City of Seattle, 2018). Durkan was elected in 2019 to the C40 Cities Steering Committee, the body that provides direction for the green global network of 94 cities, memorably noting that cities like Seattle are not just "thinking about tomorrow, but about the next 100 years" (C40 Cities, 2019).

Mayor Nickels, though, is arguably the crucial political figure in permanently shifting urban policy priorities and interinstitutional attention from wider sustainability values to the specific work of climate mitigation and adaptation. The fruit was likely ripe for picking. It was time, but Nickels's "Urban Sustainability Advisory Panel" was comprised of city officials, leaders from local businesses, representatives from environmental groups, and other key stakeholder groups (Office of Sustainability and Environment, 2002–2005). In Seattle, then, it is now less a question of green mayors than an institutional greening of the mayoral office.

Unsurprisingly, mayoral politics have influenced public administration. From the mid-1990s, local institutional capacity and carbon management cultures within city bureaucracies discernibly shifted. As in other cities, Seattle increasingly staffed its traditional line agencies to deal directly with urban sustainability goals and climate action policies. After agreeing to participate in the Cities for Climate Protection program in 1997, for example, the city council increased expenditure allowances in the adopted budgets of both Seattle City Light and the Office of Management and Planning to improve greenhouse gas auditing, access new grants, and recruit companies to the "Climate Wise Local Government Industrial Partnership" program run by the Environmental Protection Agency during the Clinton administration (City of Seattle, 1997).

On this issue we see no "council wars" with the remnants of a urban machine openly hostile to new forms of carbon action, in part because "changes in government priorities came from the neighborhoods [...] before they came into the electoral campaigns or to city hall" (Clavel, 2010, p. 2; Krueger & Agyeman, 2005). Seattle consolidated its "progressive city" (if not "radical regime") credentials largely through local green action, eventually merging with labor, identity, and race themes (Dreier, 2013).

In 2000, under Mayor Paul Shell, the city created the OSE, partly to help overcome the well-known and stubborn persistence of organizational silos. Seattle has many institutional "siloes" that nonetheless seek to work together, creating complex webs of still rather understudied "state-society-economy-nature" networks (Pacione, 2014). Personnel in many city departments deal everyday with global climate policies. This is a remarkable statement in and of itself about contemporary local government in American cities like Seattle (Pettibone, 2015). Key organizational actors include, for instance, City Light; Construction and Inspection; Economic Development; Energy Management; Housing; IT; Intergovernmental Relations; Neighborhoods, Parks, and Recreation; Planning and Development; Public Utilities; and Transportation. A comprehensive review of all activities is, therefore, impossible. Yet an indicative sense of how emerging and evolving governance arrangements spool out across Seattle is needed.

As one illustration, staff at Seattle City Light manage a range of climate-relevant programs targeting hydroelectricity, integrated resource plans, resource acquisitions, and advanced/smart metering. "Energy advisors" specifically help Seattle's residential and commercial customers to understand new energy options that promote conservation, energy efficiency, and smart energy choices. These activities nominally link with many others. The Capitol Hill EcoDistrict Project, for instance, seeks to advance City Light's solar energy goals. Managed by Capitol Hill Housing, a local community development corporation (CDC), the Capitol Hill EcoDistrict Project further involves activist groups like Seattle Food Rescue, a nonprofit organization that redistributes otherwise wasted food from businesses to charities and agencies in order to serve hungry, unhoused, and low-income individuals—and, of course, mitigate local carbon emissions along the way through food recycling (Capitol Hill Housing, 2018).

These synergies go on. The Capitol Hill EcoDistrict Project supports the local sharing economy, including the Capitol Hill Tool Library, which describes itself as a network of neighbors, businesses, and community groups dedicated to making the Capitol Hill area more sustainable (Sustainable Capitol Hill, 2020). Other sharing economy entities across Seattle, which are growing, similarly focus on "mindful consumerism" and "community building" rather than "capitalist profit." Horizon, for example, uses the "couchsurfing/Airbnb model" to deliver fees to charitable organizations who are sheltering the homeless (see http://www.horizonapp.co/?utm_source=BuiltinSeattle).

Since 2000, the OSE has sought to encourage but also to coordinate these kinds of interrelated dynamics across the city. With a 2019 budget of $7.7 million, the OSE is a tiny part of the administrative components of the local state, paling in budgetary importance compared with more traditional

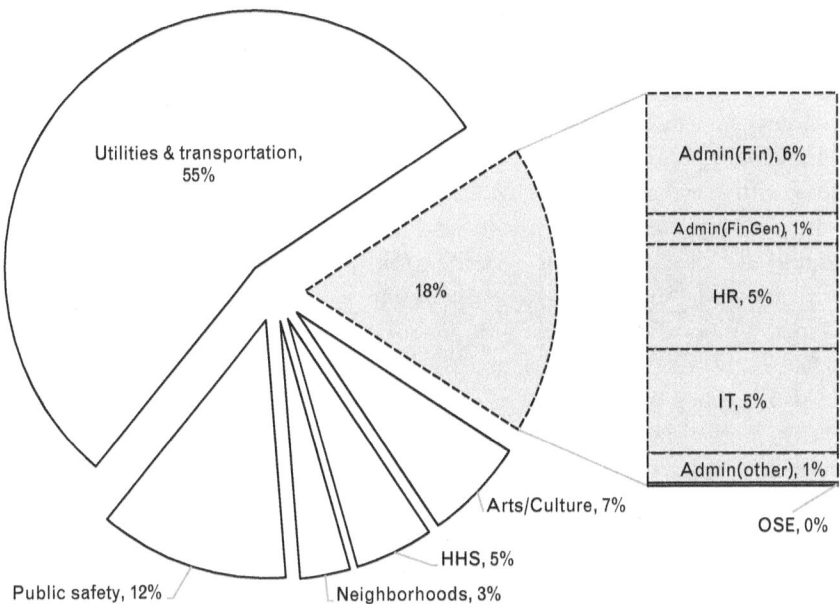

Figure 3.4. City of Seattle total budget expenditures, by area and percentage (Source: City of Seattle, 2020a, pp. 39–41)

services such as utilities, transportation, public safety, spatial planning and development support, and even other administrative services, such as information technology, human resources, and finance (Figure 3.4).

Nonetheless, the OSE synchronizes interdepartmental work around, *inter alia*, food systems, urban forestry, energy efficiency in city buildings, equity and environment, and environmental performance measurements. The OSE further conducts research and develops policies and programs, particularly around concerns with environmental equity, building energy, food policy, and transportation. Finally, the OSE coordinates the implementation and assessment of the city's CAP (Coven, 2020). Staff and advisors include specialists in citywide resource conservation strategies, environmental sustainability policy design, and, of particular importance here, climate action and climate justice.

Climate-policy expenditures pervade other institutional pieces of the pie chart in Figure 3.4. The biggest pieces are utilities (water, wastewater, solid waste, drainage, power) and transportation (which, again, is also shaped by planning and development strategies). But evidence of policy and financial change is even more widespread. In the proposed 2020 budget, the ongoing "carbonization" of urban policy includes an additional $2.2 million to continue the hydro-electrification of the city's vehicle fleet, a proposed expenditure of $600,000 by Seattle City Light to add six more direct current fast chargers for electric vehicles, and an additional $1.5 million for the city's Municipal Energy Efficiency program. In 2019, as an example that links directly with social equity goals, Mayor Durkan proposed a new tax on heating oil service providers to help low-income families convert to sustainable heating supply like electric systems. "Heating oil produces carbon pollution," her budget noted, "so the City chose this as one of its strategies in its 2013 Seattle Climate Action Plan to achieve its goal of becoming a carbon-neutral city by 2050" (City of Seattle, 2020a, p. 231).

If mitigation and adaptation policies are distinctive concepts, they nevertheless merge in carbon-neutral procedures. Mitigation policies seek to adapt various urban systems—planning, transport, energy, waste, water, housing—to a changing global climate that will, by most scientific measures, increasingly test the resiliency of Seattle's economy, society, and polity to absorb recurrent disturbances, notably flooding and heat spikes.

Like the COVID-19 pandemic of 2020, global climate change is not really a question of "if." It is not even a question of "when." It is a question of "now," as Chapter 4 will discuss in greater detail. "In this context," as Donald Nelson (2011, p. 114) maintains, "adaptation refers to the process of managing system resilience. Adaptive capacity and adaptation are the resources and processes that work to maintain the function of a system in a manner that does not lead to loss of future options." Accordingly, policy responses in Seattle reflect the steady political development of a local state facing existential realities *insufficiently addressed by federal authorities*—notably the "greening" of mayoral leadership and the "carbonization" of institutional resources and professional competencies across multiple city departments.

At the same time, interpreting these changes as features (or "traits") of Seattle *stricto sensu* misses a range of functional and policy realities that concomitantly include and stretch beyond the "city" of Seattle as a legal and bureaucratic place with defined territorial authority—a point first established in Chapter 1. Seeing Seattle-*in*-networks and regional fora means seeing various socio-technical systems as part of "local" climate action and "urban" carbon policy. Epistemologically, seeing Seattle-*in*-networks means mapping new relational geographies of teaching and learning through "trans-local" climate advocacy. Three such geographies stand out here: the national urban network of the MCPA, cofounded by Mayor Nickels in 2005; Seattle's rising position in transnational climate action networks like the Global Covenant of Mayors and C40 Cities, what I have elsewhere called a new "green internationalism" (Dierwechter, 2019); and at the metropolitan scale what I shall additionally call here a new "green city-regionalism," usefully illustrated by the K4C.

As the epigraph to this chapter notes, in other words, "climate change is not a stand-alone issue" but an ongoing process of deliberated changes that requires action from *within* the city ("local government, residents, businesses, industry, building owners, utilities, and many others") as well as from *without*, "at the state, federal, and international level" (City of Seattle, 2013a, p. 3). Both dimensions, I argue, are increasingly vital to understanding climate change efforts in, and the future(s) of, Seattle.

Seattle-in-Networks: Trans-Local Climate Advocacy

The global emergence of urban environmentalism started in the 1970s (Brand, 2005), as first discussed in Chapter 1. It took off in the 1990s, as discussed here. The International Council for Local Environmental Initiatives' (ICLEI's) Cities for Climate Protection Program, for example, has helped local governments in the United States and dozens of other countries to formulate and implement plans for reducing greenhouse gas emissions for over 30 years. In fact, ICLEI members in the Puget Sound region, including Seattle, Tacoma, Everett, Bellevue, along with King, Pierce, and Snohomish Counties, today represent one of the key "clusters" in the entire United States (Reams et al., 2012). Once again, it is not just about Seattle. That acknowledged, trans-local climate advocacy in/through Seattle took a significant step forward in 2005, when Mayor Greg Nickels launched the MCPA.

The Mayors Climate Protection Agreement: Redressing federal inaction

The MCPA was (and remains) a notable program of the US Conference of Mayors, an intercity organization that has long promoted energy and environmental conservation. A policy and advocacy network of about 1,000 mayors by the late 2010s, the MCPA has aimed since 2005 to reduce carbon emissions *in loco* below the 1990 levels as specified by the *original* Kyoto Protocol. Under the agreement, each member city strives to meet or beat CO2e targets through local policy action like land-use reforms, urban forest restoration, and public information campaigns. Cities work together to pressure their state governments, and the federal government, to enact greenhouse gas emission policies and programs focused explicitly on reduction (Dierwechter, 2010).

Michael Bloomberg's support for Nickels's original MCPA agenda during his own mayoral years in New York (2002–13) accelerated the still relatively new idea at the time that cities, especially large global cities like New York, Los

Angeles, Chicago, and to some extent Seattle, were now significant centres of power that could lobby effectively for global climate action (Curtis, 2014). As Rice (2018, p. 110) argues, the MCPA, born in Seattle, "helped propel cities to the forefront of the climate change policy debate." Still, the original motivation for the MCPA was to create a national network of "cool" governments—core cities, suburbs, counties, even Census Designated Places (CDPs)—in order to redress federal inaction during the George W. Bush administration. The key was to use existing institutional infrastructure and personal relationships built by the US Conference of Mayors.

Mayor Nickels was quoted by a reporter at the time:

> I knew that this idea would resonate in Seattle, a very environmentally conscious city, but I didn't know how it would be received in a lot of other cities. So my first outreach was to similar-minded cities like Portland [OR] and San Francisco, and then to cities like Minneapolis and Burlington [VT]. That group of mayors—there were eight or nine of us—then issued the challenge further to all the mayors. We sent a letter on March 30 to about 400 mayors. Almost immediately we started getting positive responses—some very memorable—back from cities. (Little, 2005)

Carbon politics picked up steam. Climate protection featured prominently in Seattle's 2006 state legislative agenda (City of Seattle, 2006a). A progress report issued by Seattle's OSE in 2010 later emphasized the city's redoubled efforts to reduce CO_2e emissions by building walkable communities "with climate protection in mind" through, for instance, linking "transportation choices" with more "compact communities" and also by using "clean fuels to run clean fleets" along with clean energy to support more efficient buildings (City of Seattle, 2010, pp. 4–12). Importantly, the report further stressed the need to adapt to a fast-changing global climate, which, once again, became a signature leitmotif in the landmark 2013 climate action plan.

Nickels's public entrepreneurial role in creating the MCPA indicates that Seattle has been a "leader" rather than a "follower" or "laggard" in the rise of trans-local climate advocacy networks (Wang, 2012). However, public entrepreneurship in the mayor's office should be embedded within an older and consequential history of urban environmental governance shaped by strong civic engagement and neighborhood activism—or what Sanders (2010) has documented in detail as "the roots of urban sustainability" in Seattle's "ecotopian" political culture from at least the 1960s if not earlier. While political commitments by green mayors are these days crucial in catalyzing local carbon action, as Wang (2012, p. 1119) observed in his study of the MCPA experience in California, they are "usually backed by a significant number of constituents."

Seattle's transnational climate participation

The MCPA suggests that Seattle has motivated others to govern globally from below (Sellers, 2002). The more recent diffusion of trans-local climate action networks that now span national borders—referred to in the academic literature as transnational municipal networks (TMNs)—additionally implies that Seattle is learning from new global spaces of interwoven local carbon agency. Seattle is an active member of three of the best known TMNs in the world: the aforementioned ICLEI's Cities for Climate Change, the world's oldest TMN; C40 Cities, established in London in 2005; and the recently reformulated Global Covenant of Mayors (a merger of the Covenant of Mayors, started in 2008; the Compact of Mayors, started in 2014; and Mayors Adapt, started in 2014). Seattle is a member of other global climate action networks, including the Carbon Neutral Cities Alliance, whose participants include neighboring Vancouver, British Columbia; Portland, Oregon; and San Francisco, as well as other global cities such as Copenhagen, Melbourne, and New York.

Ascertaining the precise empirical impacts of specific TMN projects on specific cities like Seattle is methodologically difficult, as they advocate and assist with similar policies and commonly merge with ongoing urban practices (Heikkinen et al., 2020). In their study of the overall influence of TMNs on 377 world cities, Heikkinen et al. (2020, p. 7) conclude that "network membership does support the [climate] adaptation process" and thus complements work on older "mitigation" policies. Specifically, they hypothesize that network members like Seattle are more advanced than nonmembers in their climate change adaptation planning, and, no less significantly, "the more networks a city is a member of, the more advanced it is in its adaptation planning" (p. 3). What evidence, then, is there for TMC influence in Seattle, for "advanced" local learning from a new type of "green internationalism" (Dierwechter, 2019)?

A fashionable area of planning and policy development is integrated and comparable data analytics, such as the Global Covenant of Mayors' "Data4Cities" initiative. This initiative involves representatives from C40 Cities, the European Commission, ICLEI, and the United Nations Environment Program (Global Covenant of Mayors, 2018, pp. 18–19). The idea is to standardize climate and energy data by streamlining how and what information is collected and analyzed, allowing members cities, in theory, to "showcase achievements while unambiguously tracking progress" (p. 3).

In practice, standardization requires member cities to develop reporting frameworks for (1) greenhouse gas emissions inventory, (2) target setting,

(3) risk and vulnerability assessment, and (4) climate action and energy access planning. Ultimately, Data4Cities shares procedures related to data collection, data management (and access), data validation, analysis, and dissemination (p. 4). With respect to adaptation planning, "risk and vulnerability assessment" include mandatory, recommended, and optional reporting. Cities identify the most significant climate hazards they face—or expect to face—from a climate change "hazards matrix" that includes: extreme precipitation events, wind and storm surges, extreme temperature spikes, growing water scarcity, wild fires, flood and sea-level rise, chemical changes (for instance, ocean acidification), mass movements (slope failures), and local-regional biological changes (water- and vector-borne diseases). Chapter 4 will return to these hazards and risks as well as related data on human systems and vulnerabilities.

According to the GCoM, Seattle had completed five of nine reporting phases by mid-2020. This record is rather similar to Portland and San Francisco and well ahead of cities like Miami and Atlanta. One of only 167 US communities to join the GCoM so far (compared with 8,839 cities in Western Europe!), Seattle's reporting procedures have benefited from ongoing, often detailed, efforts to calculate, as one example, the carbon impacts of city-owned buildings. Seattle has also conducted a "carbon intensity analysis" for major city departments by building type, aiming to reduce energy use and carbon emissions in 2025 by 40 percent as part of the city's much wider, and much harder, community-wide "carbon neutrality" goal for 2050 (City of Seattle, 2016c).

Green city-regionalism: K4C between city and suburb

Trans-local climate advocacy that redresses federal inaction through national networks like the MCPA or that signals the expansion of "cities as international actors" (Herrschel & Newman, 2017) through new spaces like the GCoM, is exciting. But it misses local dynamics of growing importance. Work on the diffusion of green policies and global carbon geopolitics through relational spaces of intercity "teaching and learning" (Lee & van de Meene, 2012) rarely highlights the *metropolitan context* within which "leading" cities engage in transnational network activities. Core (or primary) cities like Seattle, Copenhagen, Freiburg, and Portland are *de-territorialized* and too often *abstracted* from their local relationships of economic production and social reproduction (Jonas, 2020). Yet, as Mossner and Miller (2015, pp. 19–21) put it in their important critique of Freiburg, Germany, often discussed as one of Europe's greenest cities:

> Given that sustainability deals with issues of energy flows, resource flows, human mobility, and other relationships that extend beyond the

boundaries of any individual city, we ask whether it makes sense to conceptualize sustainability as a quality of discrete places. […] Freiburg has often been portrayed as an actor in charge of its own sustainability. […] Clearly there are limits to this characterization.

Seattle is not entirely an "actor in charge of its own sustainability" and, in fact, relies on a new regionalization of urban policy and climate action redolent of the geopolitical dynamics just discussed above (Moisio & Jonas, 2018). For again, "Seattle" is not only a city, as outlined in Chapter 1; it is a global city-region structured by a regional economy (Scott, 2001).

Like other places, the Seattle area is managed by formal institutions that take a regional perspective on challenges like global climate change, including the PSRC, Sound Transit, the Puget Sound Clean Air Agency, and the Northwest Seaport Alliance, the last of which is a first-of-its-kind partnership between the (long-competing) ports of Tacoma and Seattle. Both on their own and in collaboration with one another, these institutions shape city-regional governance priorities on issues like "energy flows, resource flows, human mobility, and other relationships that extend beyond the boundaries of any individual city" (Scott, 2001). Recent regional spatial and transit planning policies include an entirely new agenda for climate change. The PSRC now pays explicit attention to "substantially reducing emissions of greenhouse gases." It is thus "preparing for climate change impacts" (Puget Sound Regional Council, 2019b). For example, a PSRC draft policy for VISION 2050, MPP-CC-3, addresses "impacts to vulnerable populations and areas that have been disproportionately affected by climate change" (ibid.). As one senior regional planner recently reported, VISION 2050, the region's growth plan, now includes a more robust climate chapter with a range of new policies, goals, and actions for the metropolitan area. The PSRC is also updating its Regional Transportation Plan to advance wider climate goals. This initiative deploys a "Four-Part Greenhouse Gas Strategy, which is reevaluated every four years." VISION 2050 will furthermore develop a "climate wedge analysis," which widens the traditional focus on on-road mobile source emissions and strengthens partnerships with the Department of Ecology and the Puget Sound Clean Air Agency (McGourty, 2020).

This "greening" of regional planning and policy institutions is thus complemented by the recent development of informal, "softer," networks of climate action and coordination between (parts of) municipalities and other key actors, including the Port of Seattle. Examples include the Puget Sound Climate Preparedness Collaborative and the K4C.

K4C is discussed further here. Building on King County's "unique position to advance regional solutions to combatting climate change" (King

County–Cities Collaboration, 2020), the K4C group, cochaired by a "Senior Climate Change Specialist" and a "Climate & Long Range Senior Planner," was formed only in the mid-2010s to enhance the long-term effectiveness of local climate action. K4C asserts that outreach and joint-funding efforts can only benefit from more focused collaboration between staff, as can related efforts to share standards, benchmarks, strategies, and data. A particularly interesting feature of K4C is joint work to shape state legislation conducive to regional and local carbon action, such as more support for a price on carbon that reinvests revenues back into efforts to reduce greenhouse gas emissions (e.g., transit service, energy efficiency and renewable energy projects, forest protection and restoration) and that, furthermore, "prioritizes investments that benefit communities most impacted by climate change, and ensure a just transition for workers in fossil fuel industries" (King County–Cities Collaboration, 2020, pp. 2–4).

At the opposite end of this political and policy advocacy are K4C's various on-the-ground projects. The new network is trying to improve tree canopies, solar energy use, street lighting, green business accelerators, and so on. Such projects and programs are small-bore, often spatially and institutionally fragmented, and systemically incremental rather than structurally transformative. Yet they illustrate how "urban climate action" works in daily practice and how it often lands outside of Seattle per se. Using funds from the Washington Department of Commerce's Clean Energy Fund, Seattle City Light, and King County's Fund to Reduce Energy Demand (FRED) program, K4C has facilitated, for instance, solar energy investments in King County's parks. The combined systems generate 358,000 kWh each year, saving King County utility costs and reducing CO_2e emissions by 229 metric tons per annum. Initial investments will be paid off in the late 2020s through utility savings, tax credits, and solar production incentives (King County–Cities Climate Collaboration, 2018). These projects reveal tangible efforts to decarbonize public spaces that benefit wider populations rather than just private entities—and thus to rebuild "the public city" through greener social infrastructures of shared urbanism (Perry, 1995).

Conclusion

What should we make of Seattle's various efforts since the 1990s to build a "climate-friendly" city in an unsustainable world? In August 2019, the Seattle City Council, in political charge of one of the best-educated and most prosperous places in the contemporary United States, adopted Resolution 31895 on starting "a city-level green New Deal" (Tigue, 2019). It is worth quoting at some length:

Be it RESOLVED by the city council of the City of Seattle that:

Section 1. The City of Seattle ("City") supports policies that promote strong families and communities, including paid family and sick leave, affordable child care, universal health care, and high-quality, free educational opportunities for all as laid out by the federal Green New Deal resolution and urges the United States Congress to pass the Green New Deal.

Section 2. The City recognizes that, while it has made some progress towards reducing its dependence on fossil fuels, that progress is insufficient to make the necessary changes to shift Seattle's economy to be more equitable and ecologically sustainable.

Section 3. The City envisions a future where Seattle residents can live healthy, prosperous lives, free of toxic chemicals and fossil fuels, and the social and ecological well-being of all people is prioritized over the profit of private corporations.

Section 4. To achieve this vision, the City commits to creating a Green New Deal for Seattle, with the following goals:

A. Make Seattle free of climate pollutants, meaning those that cause shifts in climate patterns, including carbon dioxide, black carbon, methane, nitrogen oxides, and fluorinated gases, by 2030;

B. Prioritize investment in communities historically most harmed by economic, racial, and environmental injustice. (City of Seattle, 2019)

The diction here, especially in Section 4.B, is what I have elsewhere called "state-progressive" (Dierwechter, 2017), albeit influenced significantly by the more "radical-societal" neighborhood-based politics of a city long shaped by social activists and organized environmentalists (Sanders, 2010). Section 2's statement that "the social and ecological well-being of all people is prioritized over the profit of private corporations" also stands out. It is aspirational and ecotopian, certainly, and likewise candid about the limitations of what has been achieved thus far, both environmental and social—and thus political, too. Seattle "has made some progress towards reducing its dependence on fossil fuels," yet this progress, the Council fully admits, "is insufficient to make the necessary changes to shift Seattle's economy to be more equitable and ecologically sustainable."

So, some "progress" on fossil fuels and climate pollutants, but "insufficient." For radical critics of Seattle's climate action story so far, Seattle's dynamic economy—high-tech, informational, increasingly unequal, gentrified, and embedded deeply in the global production chains of "capitalism's war on the Earth" (Foster et al., 2010)—*cannot* be shifted incrementally to a more equitable and ecologically sustainable place. "Radical visions" to promote global climate responsibility through local food sovereignty in Seattle, Alkon and Mares (2012, p. 347) have specifically reasoned, are simply too constrained "by [the] broader forces of neoliberalism." For Seattle's "green" business community, in contrast, formal political attacks on "the profit of private corporations" are jarring from an American government at any level and pragmatically antithetical to how that community perceives effective climate action. Founded in 2003, for example, the Seattle-based Network for Business Innovation and Sustainability has helped to focus "mainstream" policy attention on "sustainability as a business opportunity." Similarly, the Seattle Metropolitan Chamber of Commerce has argued that "by unleashing the talents of our businesses, we will continue to invent products and services that will reduce carbon not only here but around the world" (McIntyre, 2014). Put simply, unleash markets on ecosystems through technological innovation.

Despite these differences, this chapter's focus on Seattle's policy progress since the mid-1990s suggests that the reality of climate *change* is not particularly contested, a theme I develop more fully in Chapter 4 (Pacione, 2014). It is not fake news for citizens. It is not a hoax for leaders. Evoking the prospects and perils of climate change in, around, and through Seattle's socio-technical systems—planning, housing, transport, energy, water, parks, and so on–has become what sociologists of science and technology call "strong rhetoric." Climate dissenters feel lonely. A string of Seattle's mayors has advocated (with little real resistance) for the mounting importance of, first, *mitigating* CO_2e emissions, and second, *adapting* to the world's general failure to do so. Strong rhetoric transforms an antisocial nature "out there" somewhere with its own rules into a social nature, a second nature, a produced nature, a hybrid nature (Pollini, 2013). Climate *change* is the *nature* that modern cities like Seattle make—and can, in some theories of transition, unmake. That is the fervent hope.

"Unmaking" climate change in Seattle has meant, in practice, a thousand tangible things in dozens of major policy areas that define and shape daily urban life—some big, some small: namely, public entrepreneurship; growing transit communities in urban villages and centers; shifting to light rail options; linking sidewalks together; expanding the sharing economy from experiences and services to coproduction like FabLabs and industrial ecology (McLaren & Agyeman, 2018; Janos, 2018); improving inter-port freight logistics; developing matrices of hazards and vulnerabilities, crafting new development codes (Atkinson, 2017; Vincent, 2019); accessing green funds; resisting federal stasis; building city-regional, national, and global networks to coordinate data sharing; putting solar panels in public parks; and staffing line agencies with experts in environmental sustainability policy design, equitable transit-oriented development, and climate justice—to recapitulate just a few examples briefly discussed here and thus to peer into what I previously called the micro-capillary work of *ordering* complex, heterogenous "sociotechnical systems."

That work is far from cohering into a new urban reality. Radical groups and business associations alternatively seek to co-shape the local state's internal institutional routines and investment priorities. Developmental, redistributive, and distributive public policies, moreover, reflect not only the power of market dynamics and popular control systems, but, as Paul Kantor (1988, p. 25) originally argued, intergovernmental systems as well. Yet the future is now—and it is a frightening prospect when we actually contemplate in detail the variegated implications and impacts of climate change on cities like Seattle, the subject that now follows.

Chapter 4

THE FUTURE: CLIMATE CHANGE, SOCIAL VULNERABILITIES, AND THE TRANSFORMATIONAL AGENDA

> Climate change should be studied as a hybrid condition using multiple epistemological frameworks and integrating social and biophysical sciences
>
> Burnham et al. (2016)

Contemporary climate change is the nature that cities make. This nature is a social nature, a second nature, "a hybrid condition" (Burnham et al., 2016; Pollini, 2013). Contemporary climate change is not "natural," in other words, subject simply to the internal properties of the biophysical sciences, as crucial as these sciences are to urban climate research and, as we shall see, this chapter's main themes. It is also subject to what we know from the social sciences—to the "market failures" of unpriced pollution, for instance, no less than to the "slope failures" of water-drenched ravines. As one group of biophysical researchers observe: "Greenhouse gas emissions are influenced by a wide range of complex social, political, and environmental factors (population growth, geopolitics, technological innovations, etc.) [...] [which] are difficult to predict" (Mauger, et al., 2018b, p. 10).

Contemporary climate change is the most significant socio-natural product of the Anthropocene, created from the steady carbonization of urban economies that first began in the United Kingdom in the mid-eighteenth century with the Industrial Revolution. It is largely the result of cost-minimizing enterprises with new technologies removing concentrated forms of latent energy (early on surficial coal, now various forms of deeper oil, coal, and gas) and then, decade after decade, diffusing that energy into the atmosphere though production, circulation, and consumption, where it accumulates far faster than it can descend back into oceans, soils, forests, and other "sinks." The atmosphere, no less than land, is now a geopolitical space of this second nature (Uloa, 2017).

The manifestations of accumulation—atmospheric and capitalist—are unmistakable: global warming, rising sea levels, more extreme weather patterns and events, and fast-changing ecosystem dynamics, all of which make contemporary life more vulnerable. At the risk of oversimplification and the elision of many crucial issues, these manifestations and vulnerabilities in Seattle are disproportionately related, in various ways, to the unjust impacts of water—a thematic focus first broached in Chapter 1 and developed further here. "One of the most serious impacts of climate change is how it will affect water resources around the world," the Canadian environmentalist David Suzuki notes, "[for] water is intimately tied to other resource and social issues such as food supply, health, industry, transportation and ecosystem integrity" (David Suzuki Foundation, 2020).

Using projections and ideas from the ARC3.2 report as well as local interviews with experts and especially research conducted by local scientists over the past several years, the first part of this chapter speculates on the most likely "climate scenario" for the Seattle area in the coming decades of this century, again organized especially around the synoptic theme of water.

Imagining a future climate "setting" much more like contemporary San Francisco or, in the warmest models, even Southern California, the chapter then focuses in the second part on questions of social vulnerabilities in and around Seattle. Finally, the chapter considers what the city and its partners will most need to do to prepare for such a dramatic transformation and particularly the question of how likely (or not) political and policy actors will be able to meet the local challenges of global climate change. Building on Chapters 2 and 3, the discussion draws empirically on the insights of local informants and specialists in urban climate science and policy advocacy while also placing key arguments within the conceptual context of the ARC3.2 framework introduced originally in Chapter 1.

Central to these arguments, as in previous chapters, is how projected climate changes in Seattle in the coming decades will disproportionately impact disadvantaged populations without substantial political and policy efforts in multiple arenas of urban development to alleviate these impacts.

The Next Hundred Years: Twenty-First Century Climate Change in Seattle

When the mayor of Seattle commented in 2019 that the city would have to think not just "about tomorrow, but about the next 100 years" (C40 Cities, 2019), she perhaps had in mind climate change projections by C40 Cities and its global partners, including the Urban Climate Change Research Network (UCCRN) (Rosenzweig et al., 2018). According to the *Second Assessment Report*

Table 4.1. Climate Change Projections for Seattle, Washington

Base decade	Temperature (30 years out, °C)		Precipitation (30 years out, %)		Sea level (10 years out, cm)	
	Low	High	Low	High	Low	High
2020s*	0.9	1.8	−1	12	4	18
2050s*	1.7	3.6	1	13	14	56
2080s*	2.2	5.6	2	19	21	118
2100**	"Absolute" sea-level change: 2.0 (50%), 1.5–3.8 (83–17%)					

Source: *Rosenzweig et al. (2018, Appendix 2); **Miller et al. (2018, p. 14).

for Climate Change and Cities (ARC3.2) by the UCCRN, Seattle's mean temperature will rise between 2.2°C (3.96°F) to 5.6°C (10.6°F) by the early twenty-second century. Mean annual precipitation will increase between 2 percent and 19 percent. Local sea-level rise will increase between 21 cm (0.68 feet) and 118 cm (3.87 feet) by the 2090s, with local researchers suggesting a 50 percent probability of 2.0 feet by 2100 (Table 4.1).

That Seattle has a chance to be (on average) 4–10°F hotter, up to one-fifth wetter than it is now, and further under water—between two to three feet if higher projections play out and much more than that under less likely but still possible "worst case" models—is a didactic triptych of troubles to hang on the walls of elected officials (Miller et al., 2018).

As first discussed in Chapter 2, it means rainy, lush Seattle, hewed out of a temperate rain forest, and tied closely to the inland Salish Sea will paradoxically have to develop an entirely new relationship with water. Other cities—for example, Tucson, Arizona; Nanjing, China; Tacloban, Philippines; Oujda, Morocco—will likely be far drier in 2100. Not so Seattle. Life-giving and culture-shaping, water is heavy, unevenly delivered, often threatening—and more of these liquid realities are on their way in the coming decades. With luck and foresight, future mitigation efforts around the world may "bend the curve" to the lower projections given above, but significant adaptation of some kind is now inevitable, especially for coastal cities. At the same time, the COVID-19 response globally in 2020 does not assuage many fears that overall climate change preparedness in most places is remotely sufficient. And again, securing sustainability is a political, no less than technological, challenge (Laurian et al., 2017).

Temperature and precipitation figures are subject to probability ranges. The exact amount of overall sea-level rise is also unknown (Mauger et al., 2018a; Miller et al., 2018). Still, future sea-level rise is certain because it is already happening. Global levels have risen about 15 cm (six inches) since the mid-twentieth century and about eight inches along Puget Sound since 1900 (Mauger et al., 2018a, p. 11). In recent years, global sea levels have risen faster than in the past—an average of 2.5 cm (1 inch) every five years, a trend that scientists expect to continue and perhaps accelerate. Indeed, as a recent National Oceanic and Atmospheric Administration technical report warns, recent evidence of accelerated ice loss from Antarctica and Greenland "only strengthens an argument for considering worst-case scenarios in coastal risk management" (Center for Operational Oceanographic Products and Services, 2017, p. 14).

"Downscaling" recent research on twenty-first century climate change in Seattle and the Puget Sound bioregion has focused on various aspects of overall temperature change, rising yet also more uneven precipitation, and higher sea levels—as well as related concerns with local water and heat stress and drought, the greater intensity of storm surges and regional flooding, local air quality, ocean acidification, snowpack delivery, biosystem integrity and coastal upwelling, and changes in the spread of infectious diseases (UW Climate Impacts Group, UW Department of Environmental and Occupational Health Sciences, Front and Centered, & Urban@UW, 2018). Research thus considers a number of potential direct impacts, such as storm surges and extreme heat stress (Jackson et al., 2010), as well as indirect impacts, such as coastal and estuarine fishery ecosystem deterioration (Reum et al., 2011; Baker, 2020).

Realizations vary because "the climate system is chaotic," so that small changes in weather climate variability can make big differences over time (Mauger, et al., 2018a). Tectonic activity also matters over time, especially if/when Seattle is hit by a large earthquake. Moreover, regional climate models (RCMs) that simulate local-scale climate changes are hard to build and still uncommon (Salathé et al., 2010). An inland sea like Puget Sound that is surrounded by significant mountains on two sides constitutes a complex orographic terrain that differentially interacts with incoming "atmospheric rivers." This presents particular challenges for precise downscaling to a specific place like Seattle; perhaps one-third of "extreme precipitation events" in one region, such as the Olympic Mountains to Seattle's west, may not actually occur in Seattle, and vice versa (Lorente-Plazas et al., 2018).

The ongoing scientific search for pinpoint precision at particular places for long-range planning purposes underscores a larger concern, one based on the supposition that emissions will not be abated sufficiently in the near term. Atmospheric accumulation of CO_2e is now projected to range from 520–7,200 ppm ("low" estimate) to a more disturbing 720–1000 ppm ("moderate" estimate) by 2100. Worst case scenarios top 1000 ppm, 2.5 times higher than the 400 ppm that scientists now consider the "Rubicon" line in global warming models (Miller et al., 2018). "Future heavy rain events," Mauger et al. (2018b, pp. 1, 11) modestly conclude, "will be more intense in the Pacific Northwest." Just how "intense" will depend globally upon the amount of oceanic thermal expansion and changes in the size of Antarctic and Greenland ice sheets—as well as *in loco* infrastructural and policy efforts to alleviate the worst effects during future "extreme precipitation events" (ibid.) and other developments that nakedly expose the variegated nature of the metropolitan region's sociospatial vulnerabilities and institutional (in)capacities.

Socio-Spatial Vulnerabilities: People, Institutions, Water

For all its overwhelming and even apocalyptic qualities, contemporary climate change in Seattle and around the Puget Sound area—change in local temperature regimes, change in local precipitation patterns, change in local sea levels, change in a host of ancillary effects—will not "arrive" evenly on everyone's doorsteps. Paradoxically, rain does not fall or flow in the same ways. It does not level society's rich, middle class, and poor with the same universal forces of nature. Some people in and around Seattle are more vulnerable than others—more likely to suffer, more likely to die.

Groups like Seattle's fast-growing homeless population do not have a doorstep at all (US Department of Housing and Urban Development, 2018), a point the Seattle's "Raging Grannies" have actively protested: "With so many homes in our city | So few for the poor | it's a pity | It's time for a change | We must arrange | Homes for all!" As Ramin and Svoboda (2009, p. 660) broadly observe, homeless individuals have greater exposure and poorer protection from the elements and are more likely to occupy high-risk urban areas. Accordingly, they are likely to suffer from greater rates of illness and death associated with heat waves, air pollution, storms and floods, and vector-borne diseases resulting from climate change.

Analogies might help. The COVID-19 pandemic in 2020 showed early on that "black and Hispanic Americans were dying at rates far higher than white Americans" (NYT Editorial Board, 2020). A "hybrid," socio-natural condition, contemporary climate change is today—and will likely be in the future—no less unjust, unfair, or unkind, exposing as the COVID-19 pandemic did in 2020 and thereafter, all the extant societal tensions, institutional rigidities, and systemic biases as the biophysics and chemistries of wind, heat, and water impact local communities. Proactively identifying and mitigating these tensions, rigidities, and biases is part and parcel of how cities like Seattle will need to adapt going forward.

Several recent developments in Seattle and the region provide insights into how climate change and social justice will intersect pragmatically at the scale of urban policymaking and service delivery.

The city of Seattle's Environmental Justice and Service Equity Division, as a first example, advances the multi-departmental Race and Social Justice Initiative through a Racial Equity Toolkit. The Race and Social Justice Initiative was started in 2005 under Mayor Mike McGinn. It continues to shape municipal approaches to governance, including staffing and organization, community outreach, and program evaluation protocols (Seattle Office for Civil Rights, 2008). As discussed briefly in Chapter 2, Seattle Public Utilities (SPU) now uses this toolkit to deliver more inclusive and equitable services and especially to try to avoid unintended impacts on low-income communities of color.

Similarly, the Office of Sustainability and Environment (OSE) uses an "Equity and Environment Agenda" to foreground both race and social justice dimensions of urban environmental and climate action policies. One recent report quoted an East African immigrant who drives to a better neighborhood to take a healthy walk: "The sidewalk does not exist in my neighborhood and has poor pavement" (Office of Sustainability and Environment, 2016a, p. 10). Making African immigrants drive in order to walk is exceptionally "fine-grained" qualitative information compared with large-scale climate modeling, but it captures the contradictory patterns of everyday life in Seattle that must change (and see Raven et al., 2018).

These patterns are evident in a host of other areas. One census tract analysis across Seattle shows that low-income and minority households enjoy less tree canopy overall, "creating a racial/ethnic and socioeconomic disparity in exposure to high temperatures that [are otherwise] ameliorated by shade" (UW Climate Impacts Group et al., 2018, p. 26). Wealthier single-family residential areas benefit from 63 percent of the city's canopy cover, whereas poorer multifamily areas experience only 9 percent of this cover and thus more intense "heat island" effects that will worsen with climate change absent corrective mitigation efforts over time (Office of Sustainability and Environment, 2016b).

Two leading climate justice organizations in Seattle—GotGreen! and Puget Sound Sage—have further elaborated upon this concern:

> Seattle ranks among the top ten cities in the US with measurable heat island impacts, with a 4.1 degree urban/rural temperature difference. Due to a legacy of institutionalized racism and classism leading to a higher prevalence of pre-existing health conditions, poorer quality health care, lower building quality and fewer resources to respond, Seattle's low-income neighborhoods and communities of color are disproportionately vulnerable to heat waves, which will become more frequent as the climate changes. Further, because of our historically cool

weather, much of our housing stock does not have air conditioning—creating more risk for low-income people who don't have access to effective cooling systems. (GotGreen! & Puget Sound Sage, 2016, p. 28)

Another urban-environmental and planning advocacy organization based in Seattle, Futurewise, has created a "Climate Challenges Atlas" that uses simple bivariate choropleth maps of census tracts to show various kinds of socio-spatial "overlap" between Seattle's "historically marginalized communities," on the one hand, and the "expected impacts" of multiple climate change issues, on the other (Futurewise, 2017).

Futurewise's atlas of Seattle's "gentrifying riskscape" (Abel et al., 2015) illustrates highly uneven socio-spatial patterns of future vulnerability to nascent climate change patterns associated concretely with, *inter alia*, urban heat effects; poorer air quality; flooding events along with untreated stormwater discharges and combined sewer overflows; landslide impacts on public transit systems; reduced access to traditional foods, such as fish, shellfish, and native plants; and thus, especially when combined, disproportionate impacts on the physical and mental health of Seattle's low-income populations of color, including the third highest absolute concentration of homeless individuals (and families) in metropolitan America (US Department of Housing and Urban Development, 2018) (Figure 4.1).

Researchers working in interdisciplinary teams have explored the combined effects of local heat events and worsening air pollution for the Seattle city-region on future mortality rates, especially for the elderly. Jackson et al. (2010), for example, provide various models of the projected growth in mortality rates for nontraumatic and cardiopulmonary causes within King County due directly to increases in ozone concentrations of CO_2e. This research builds on previous work in urban studies that shows how the social and economic factors just discussed influence mortality rates during periods of excessive heat, such as the Chicago Heat Wave in 1995 (Browning et al., 2006). As in Chicago, the uneven impacts of future climate change on real people in real communities are already occurring in Seattle, too. Isaksen et al. (2015) have documented increased and punctuated hospital admissions associated with extreme-heat exposure within King County between 1990 and 2010. These trends will continue to grow more important and worrisome.

Hospitals and public health agencies are palpable institutions that individuals and families rely on during periods of sudden vulnerability. Less palpable but arguably no less critical are a huge range of other institutions, public and private, that underpin and help to reproduce contemporary urban society, from efficient and safe transit lines to sewerage, waste, and stormwater systems. All of these are now—and will increasingly be—under mounting pressure as

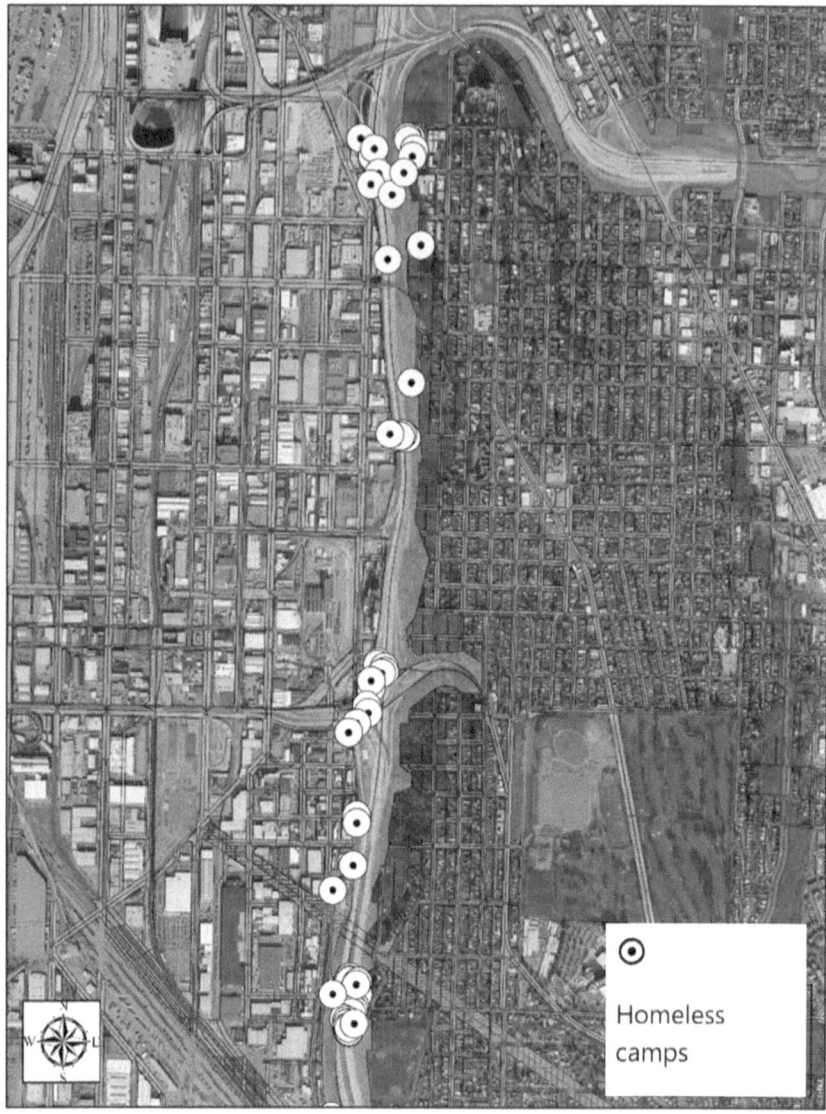

Figure 4.1. Homeless encampments along the I-5 Corridor in Seattle (2018) (Source: Author, rendered from City of Seattle, Seattle GeoData)

climate change places novel demands on more equitable and effective service provision.

Many of these systems, particularly bulk infrastructure outlays built in the mid-twentieth century, are aging out rapidly and reflect years, even decades, of profound underinvestment (Kessler, 2011). The widespread "decay" of urban infrastructure systems—which are disproportionately experienced by different races, places, and classes—has combined with the related degradation of the overall environment and the specific loss of associated (often uncosted) "ecosystem services," such as regulation of floods, soil erosion, and disease outbreaks (Doyle et al., 2008). The American Society of Civil Engineers (ASCE) gives the infrastructure of the United States an overall D+ grade; Washington gets a C grade, with the main concerns around fraying transport systems.

Efforts to redress systemic aging, mitigate carbon impacts, and adapt to emerging realities like more intense rainfall, storm surges, and higher sea levels are evident in the recent policy history and plans of SPU. As mentioned earlier in this chapter, SPU uses the Racial Equity Toolkit. In addition, SPU is a superb example of a "city" institution that transgress scales and territories of power, pushing how we imagine "urban" change through relational spaces of regional resource development and climate change management. Finally, as discussed in Chapter 3, SPU is one of the core components of the local state with the biggest budgets; it is one of the biggest contributors to climate policy and urban resiliency. Reflecting in more detail on SPU is instructive in the context of the main themes of this chapter and the book as a whole.

Seattle Public Utilities: Mitigating vulnerabilities, adapting creatively?

SPU, created from the consolidation of the Seattle Water Department with other functions in 1997, today provides water, sewer, drainage, and waste management services to more than half the residents of King County, including the city of Seattle (Haskins et al., 2002). It is one of the largest such institutions in the Pacific Northwest, employing 1,400 people who administer an annual total budget of just over $1 billion. About 70 percent of SPU's workforce is represented by organized labor.

As discussed initially in Chapter 2, pioneer Seattle constructed public sewers for rudimentary wastewater in the late 1880s, which soon included the development of the Cedar River. With postwar growth, the city added the South Fork of the Tolt River to the water supply system in 1964 (see Figure 1.2). In the 1980s, sewerage services expanded to include better coordinated drainage management and planning, with attendant user fees. Other than residential and commercial garbage collection, SPU's main lines of business today are drinking water provision and drainage/wastewater management. The decades-long development of SPU is thus about the technological expansion of the "city" into the wider bioregions and drainage basins of Western Washington as well as the politico-economic shift from a world of uncoordinated private vendors to an integrated metropolitan-wide government traditionally associated with the "natural monopoly" features of a public utility agency.

For the past 30 years, SPU's policy and project concerns with climate change have grown, with respect to both how it mitigates the carbon effects of its own infrastructural systems and how it equitably and efficiently adapts these systems to global and regional climate change projections. According to Valerie Pacino (2014) of the OSE, SPU has "a high level of expertise in-house and have been at the leading edge of this [climate] work for a long time, [whereas] other departments have just begun considering projected climate impacts." SPU has been studying the possible impacts of future climate change since the early 1990s (Seattle Public Utilities, 2007, pp. 1–14). Starting in the early 2000s, SPU commissioned a series of local research projects with the University of Washington, whose Climate Impacts Group has published over 730 scientific articles on climate change issues since 1995 (see https://cig.uw.edu/). SPU officials subsequently began to work with various other networks, including: the Northwest Regional Modeling Consortium; the Climate Impacts Research Consortium; the European Union–funded project, PREPARED, which focused on improved provision of water supply and sanitation; and the Water Utility Climate Alliance. Outputs from these groups and other initiatives influenced the original design, improvement, and implementation of SPU's 2007 Water Systems Plan, for example, which included discussions of the potential impacts of climate change on local water supplies and emerging water quality issues as well as concerns with aging infrastructure and service standards. These in turn informed race and equity themes now becoming more prominent in all urban policy arenas across the city (Seattle Public Utilities, 2007, pp. 1–15).

Water concerns, however, merge with longer-running rules, policy commitments and institutional routines that continue to shape SPU's contemporary activities, suggesting the intercurrence of multiple orders within this single agency. Taken together, SPU's temporally variegated laws, rules, regulations, programs, and plans over many decades nonetheless have "enabled resilience" in Seattle's water services, as Inha and Hukka (2019) argue. They point out that because Seattle depends upon consistent precipitation and snowpack flows into the Green River and Told River watersheds, respectively, "climate change is a considerable stressor in the region" (p. 97). Combining mitigation and adaptation efforts to confront hazards and future risks, in their view, *actively builds resiliency*, which they define simply as the ability to "recover from disturbances."

While some might emphasize techno-engineering solutions, about which more below, Inha and Hukka instead excavate SPU's institutional, policy, and governance history. Key "development steps," as they call them, include the public purchase of watershed lands in the 1920s, for instance, which continues to matter for water quality in an era often ahistorically essentialized as "neoliberal" as well as the *original* decision by Seattle's mayor in 1880 to build a gravity-supply system from the Cedar River. With respect to managing wastewater, Inha and Hukka further note the ongoing influence of a 1958 policy shift associated with the National Pollutant Discharge Elimination system and the 1972 Clean Water Act, which "Federalized" key aspects of local water management. Voter approval for a regional, watershed-based wastewater treatment system—Metropolitan Municipality of Seattle, first discussed in Chapter 2—was also a key moment in building contemporary climate resiliency. The past, then, is not really "in the past" nor always a burden. It can provide institutional foundations upon which to build innovative and embrace new trends, albeit not without tensions and barriers to transition (Haskins et al., 2002; Tackett, 2008).

This is seen in SPU's stormwater policies and practices. As climate change raises local temperature regimes (see Table 4.1), more water will likely come as heavier winter rains than slow-melting summer snow, forcing shifts in local management strategies. The region's first-ever Comprehensive Drainage Plan in 1988, which included stormwater fees, and, in 2009, new stormwater codes nonetheless have helped to support recent goals and landscape reforms set out in SPU's 2013 green infrastructure program.

The green infrastructure program builds on the accumulation of *in situ* knowledge associated with low-impact development practices across the region for about 20 years (Johnson & Staeheli, 2006). This includes technical knowledge, such as the performance of native soil infiltration rates, as well as governance and management experience. Tackett (2008, p. 320), for instance, notes that "barriers" to the rapid diffusion of low-impact development techniques and the more widespread adoption of natural drainage systems across the Seattle city-region are "traditional land use codes, including typically wider streets, curb-and-gutter, and piped stormwater infrastructure," that is, "modernist" planning theories conceived of decades that still strongly shape social-natural spaces (Whittemore, 2015).

Working through these barriers—these manifestations of institutional intercurrence—are small forms of "sustainability transition," although some forms of transition are easier and more superficial than others, and *where* they occur is an important dimension of *what* they mean (Coenen et al., 2012).

Two projects illustrate this argument. An early natural drainage retrofit project in Seattle, Street Edge Alternatives ("SEA Streets"), successfully negotiated new street design standards between SPU, focused on water management, and the Seattle Department of Transportation, focused on facilitating vehicular mobility, for a single street in an otherwise car-oriented residential neighborhood in Northeast Seattle. "Post-retrofit" runoff decreased nearly 100 percent, aiding the health of the Piper's Creek Watershed, which in turn runs directly into Puget Sound (Tackett, 2008). But such retrofits do little for racial equity and social inclusion, and may amplify urban inequality by adding public value to private homes in well-established neighborhoods—a process of "decoupling equity from environment" also seen in high-tech cities like Austin (Tretter & Mueller, 2018).

In his comparative analysis of the "politics of urban run-off" in Seattle and Austin, Karvonen (2011) has argued that, after a long period of conquering nature, in recent decades Seattle has been trying to redeem (and deepen) the relational, socio-natural ties that coshape what the city has branded as its "metronatural" identity. Karvonen pushes this idea further with far more ambitious calls for "a civic politics that develop experimental projects of relation building" (p. 194). Here experiments are not only social but socio-natural, bringing about new urban forms, certainly, but also new types of citizens—people who care about each other and about nonhuman species and in their common ecosystems.

More interesting, then, are SPU's efforts to transition toward Julian Agyeman's "just sustainabilities" as they merge green infrastructure goals with, for instance, affordable housing projects. The exemplary case here is probably still High Point, a subsidized public housing (HOPE VI) development in West Seattle managed by the Seattle Housing Authority, which itself pioneered the country's first racially inclusive public housing in the 1940s when cities like Chicago and Boston did precisely the opposite. High Point was and remains one of the largest urban applications of natural drainage systems in the United States (Tackett, 2008). With a "walkscore" of 91/100, daily errands in High Point do not require a car, while featuring extensive use of bioretention and conveyance swales, pervious paving, downspout disconnects, rain gardens, and extensive tree preservation (Figure 4.2). Here the smartness discourses of New Urbanist design ideals and parallel arguments around deconcentrating poverty meet up with green infrastructure and nonlocal climate responsibilities.

For all the fashionable talk of public–private partnerships and the steady march of neoliberal governance, also critical, I argue, are *public–public* partnerships that, here and there if not everywhere nor at all times, break down the clichéd "siloes" of public power and bureaucratic stasis, at least when

THE FUTURE 77

Figure 4.2. High Point's low-impact development, West Seattle (Source: Author).

informed by institutional legacies that carry democratic (rather than market) values of racial inclusion and social equity into climate projects and ecological programs (Oseland, 2019). All the same, as Karvonen (2011, p. 193) further notes, "technomanagerial culture" is no less resistant to "change"—even to modest, incremental transition, much less more systemic transformation—than the "obduracy of urban infrastructure." This reality points us away from considering only the possibility of future changes and toward assessing the probabilities of urban transformation as well. That in turn requires us to look back across both this and previous book chapters as we ultimately imagine various "futures" for—and in—Seattle.

Assessing the Probability of Urban Transformation

Claiming that contemporary climate change is a "hybrid condition" does not mean that climate change is easily "unmade." Specifically, it does not mean that society has endless time to blunt (much less reverse) the power of biophysical processes like atmospheric carbon loading to overwhelm *in toto* the (hopefully robust) human agency of urban policies, regional networks, national programs, or global diplomacy. That sounds alarmist, but by all accounts, political (and maybe human) time is running out in the face of planetary urbanization. Species exist as "assemblages," and their degradation may be less a downward "slope" than a sharp "cliff" at some point this century (Trisos et al., 2020). Dealing with slopes is clearly far preferable to dealing with cliffs. Peter Taylor et al. (2020) put it this way in their elegiac treatise on the contemporary climate emergency:

> Having identified anthropogenic climate change as a complex problem it must be added that from humanity's perspective it is unique. It is qualitatively different from all other organized complexities scientists study and make policies on because for humanity it represents an existential threat. Take a moment to think about that. It is simply incredible that we should be writing about such a threat. The adjective, "existential," means the readers' grandchildren, great-grandchildren or great-great-grandchildren literally will not exist. Just having to state this chilling possibility is shocking beyond belief. (p. 4)

Following the ARC3.2 framework, Chapter 1 first suggested that cities like Seattle can contribute to climate change mitigation and adaptation and thus bolster local resiliency and risk profiles by pursuing five major "pathways of urban transformation" (Figure 1.1). These are briefly reiterated here as: integrating mitigation with adaptation; coordinating disaster risk reduction

(DRR) with climate change adaptation (CCA); cogenerating risk information; focusing on disadvantaged populations; and advancing governance, finance, and knowledge networks.

Each of these "pathways" is worth a whole book, and even then, a comprehensive audit in the Seattle city-region would not be possible. Instead, Chapters 2 and 3 (as well as this chapter) have provided selective and hopefully indicative examples of various climate change efforts in and around Seattle. Taken together, these efforts—and additional examples given below—do not suggest a singular, integrated, "future" but more likely multiple futures, some more comforting than others. Different institutional pathways progress at different speeds. They show different policy and project geographies as they abut and grate with one another. They project the multiple orders of past developments, current knowledge capabilities, and ongoing cultures of governance. "Cross-cutting," multi-institutional, cross-scalar initiatives to integrate, plan, make more equitable, and finance these pathways simply seek to overcome these challenges.

Integrating mitigation with adaptation

The ARC3.2 discussion emphasizes the need to integrate mitigation with adaptation, providing several conceptual tools that are analytically helpful in clarifying and assessing Seattle's climate change efforts over the past several decades (Grafakos et al., 2018). Albeit my own interpretation and selective application of ARC3.2, Figure 4.3 places some of these conceptual tools within the context of the specific themes and issues discussed above and in previous chapters for Seattle.

Closely interlinked in climate change theory and applied urban practice, mitigation and adaptation are nonetheless different. From a temporal perspective, mitigation measures, such as Seattle's efforts to switch municipal fleets to green fuels, are long-term in focus; adaptation measures, such as strengthening flood defenses through green infrastructure, play out over shorter time horizons (the next storm). From a scalar perspective, mitigation measures are global, as CO_2e emissions impact a shared atmospheric space; adaptation measures typically target local-regional efforts to improve neighborhood resiliency in the face of multiple hazards and risks associated with global climate change (Grafakos et al., 2018).

The "zone" of policy and programmatic integration between mitigation and adaptation measures, respectively, is shaped by three socio-political and techno-managerial processes of carbon governance: maximizing synergies, minimizing conflicts, and managing tradeoffs. In (and around) Seattle, global mitigation efforts inform local adaptation through various networks, while the

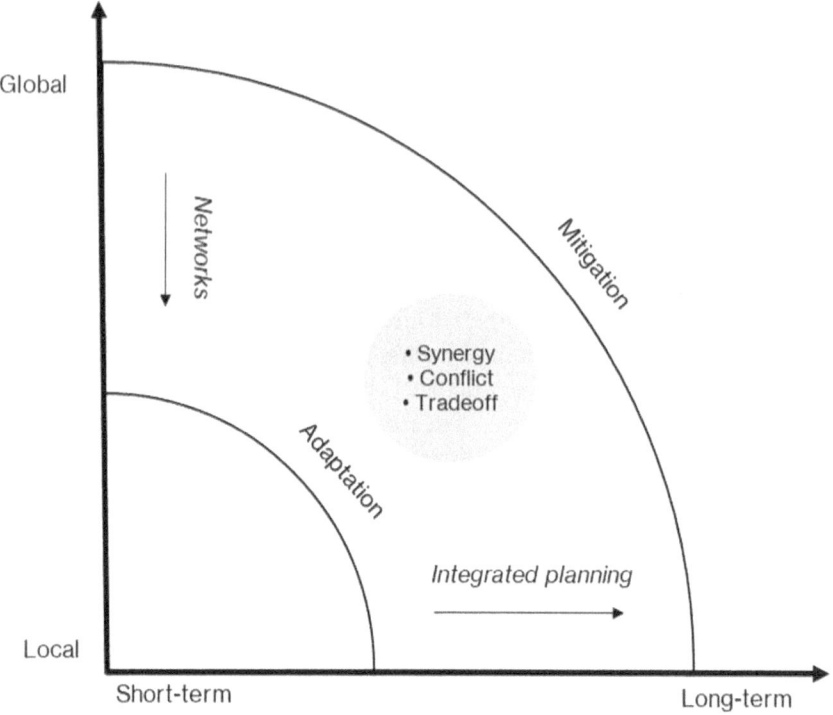

Figure 4.3. Integrating mitigation with adaptation in Seattle (Source: Adapted from ideas developed by Grafakos et al. (2018))

integration of local adaption projects into (older) mitigation measures occurs through integrated planning, particularly but not only via the climate action plans discussed in Chapter 3.

"Integrated planning" is easy to invoke yet hard to measure. Integrated planning is not the strategic implementation of a single all-encompassing "climate action plan" per se but the myriad ways in which an ecology of different kinds of plans, budgetary outlays, and institutional routines work together to change a city and a citizenry over time—sector by sector—which in turn requires constant "silo" transcendence and institutional capacity.

The Department of Finance and Administrative Services (2019), for instance, observes that Seattle has "acutely focused" on reducing fleet emissions "in support of the City's [2013] Climate Action Plan" through, *inter alia*, fleet electrification, reduced fuel use overall, and the greater use of non–fossil free fuels (p. 3). Fortunately, fleet electrification in Seattle benefits enormously from Seattle City Light, which became the first electric utility in the United States to achieve zero net GHG emissions in 2005. That said, Seattle must still develop and maintain a charging infrastructure network, which means coordinating with various city departments and private firms to install infrastructure appropriately, maintain charging stations, and identify locations for backup generators required for emergency response—among other management challenges. Between 2013 and 2019, Seattle reduced its total municipal fleet kgCO2e emissions by 12 percent and its kgCO2e emissions per vehicle by 16 percent (p. 15). Accomplishing this reduction, while also indicating infrastructural concerns for locating and maintaining "backup generators required for emergency response," suggests small but important "synergies" between mitigation and adaptation efforts in Seattle.

"Conflicts" and "tradeoffs" nonetheless pervade efforts to integrate adaptation efforts with mitigation policies (Grafakos et al., 2018). Synergistic policies—creating effects greater than constituent components (p. 108)—can concomitantly expose these conflicts and tradeoffs to the difficult politics of urban development. Green rooftops famously help absorb solar energy in (hotter) summers and reduce heat loss in (wetter) winter but also add to the costs of development, an important concern in a high-tech city struggling with housing affordability and homelessness (Gibson, 2004). Seattle's decades-long (and successful) efforts to forge greater urban density, which is necessary to support transit use, exposes citizens to pandemics that are expected to increase with climate change. Absent strong state intervention as in Sweden or Denmark, transit-oriented developments tend to support "green cities" but not "blue collars" (Dierwechter & Pendras, 2020), "elite enclaves" rather than "progressive beachheads"—thus producing contradictory new geographies of smart segregation and ecological gentrification even as they do help to mitigate CO2e emissions from the city and region (Dierwechter, 2013, 2017).

Coordinating disaster risk reduction with climate change adaptation

Contact and interaction between policy and advocacy communities focused on CCA and DRR, respectively, has been growing for many years. Yet as Solecki et al. (2011) point out, their integration in most communities is still substantially underdeveloped, even in cities like Seattle with relatively serious and sustained political and policy commitments to climate change. Like other cities, Seattle could benefit from "better recognition of the linkages and distinction between CCA strategies and DRR" (p. 138).

Disaster comes in many forms, but mainly as biological, geological, and hydrometeorological "events." A biological pandemic, the COVID-19 crisis of 2020 has been a profound stress test for disaster preparedness and response capabilities for governments at all territorial scales. As one of the early COVID-19 hotspots, the Seattle region appears to have fared better than the federal government. This performance is unsurprising given that the city of Seattle ranks as one of the "best prepared" cities in the United States for "climate-related hazards" according to one recent survey (Notre Dame Global Adaptation Initiative, 2018).

In addition to biological disasters like pandemics, tectonically active Seattle must prepare for geological disasters (earthquake, tsunami, volcanic lahars) as well as hydrometeorological disasters. Earthquakes can cause and have caused profound local destruction—with far-reaching urban consequences. Even the relatively mild 2001 earthquake sufficiently damaged the city's viaduct that it had to be torn down, leading to years of planning debates about the postmodern redevelopment of Seattle's waterfront along Eliott Bay (Ramsey, 2009). Climate change in Seattle especially impacts risks and hazards associated with hydrometeorological disasters, while also amplifying the biological effects of disease vectors like viruses.

Conceptualized as an abstract formula, Seattle's risk profile (R) is a product of hazards (H) and vulnerabilities (V) divided by overall capacity (C). Local DRR, then, is not only about increasing strategic capacity and planning efficacy through "potential synergies" between CCA and DDR (ibid.). Hazards like storm surges, flooding, slope failures, and summer urban heat islands are unnecessarily multiplied by the kinds of social vulnerabilities discussed throughout this chapter, even when mitigation efforts and capacity building are otherwise advancing (Shaw et al., 2010, p. 5).

$$R = \left\{ \frac{H * V}{C} \right\}$$

The lead agency in Seattle for DRR, the Department of Emergency Management, currently plans for earthquakes, landslides, liquefaction, power outages, cyberattacks, mass transportation incidents, and "social unrest" associated with these and other crises. Linkages with CCA and related efforts are evident (see, e.g., City of Seattle, 2020b), but arguably they are less developed than one might hope. Though "climate change" is frequently mentioned, the Department of Emergency Management's planning for climate change is a hotlink to the Office and Sustainability and Environment. In part, this is because climate change is not an "event" but a "condition," a long-term disease to manage rather than an acute emergency to solve.

In 2017, however, the OEM employed tools and mechanisms associated traditionally with standard emergency operations and logistics to address growing vulnerabilities in unauthorized homeless encampments across the city, suggesting new roles for emergency management in the coming years. "Inter-agency coordination" between the OEM and other key service agencies focused on finding safer living spaces, refuse collection, social services and housing, and mitigating environmental health effects in the most hazardous encampments (City of Seattle, 2017a).

Cogenerating risk information

The ARC3.2 framework further hypothesizes that, in addition to integrating mitigation with adaptation and coordinating CCA with DRR, respectively, the synoptic goal of urban transformation is more likely when "a full range of stakeholders and scientists" cogenerate both risk assessments and climate plans. Specifically, the framework claims, carbon governance processes that are "inclusive, transparent, participatory, multisectoral, multijurisdictional, and inter-disciplinary [...] enhance relevance, flexibility, and legitimacy" (Rosenzweig et al., 2018, p. 19). These phrases are heuristics for a more demanding theory of collaborative planning that is itself an idealized version of what should happen in cities like Seattle around climate change (Healey, 2006). Concrete examples of such heuristics in Seattle are easy enough to find but do not necessarily amount to a transformative urban process.

Like most cities, multisectoral and interdisciplinary planning processes in Seattle are increasingly common, notwithstanding opportunities for improved interactions between, as just discussed, CCA and DRR. As also earlier addressed, SPU started to engage local climate scientists at the University of Washington in the early 2000s, aiming to downscale global climate models in order to sharpen stormwater planning concerns and other utility risks and local hazards. To a remarkable extent, moreover, carbon measurements are progressively shaping urban politics, institutional goals, and policy design programs in cities like Seattle—a point elaborated by Jennifer Rice (2018) in her critical work on new "carbon territories" and modes of urban governance, and as examined originally in Chapter 3.

The challenge, though, is how "participatory," "inclusive" processes, especially in "multijurisdictional" settings, can augment the instrumental rationality encoded in formal modes of planning policy intelligence. In his work on CCA, Mark Pelling (2011) recounts the scientific impacts of the International Panel on Climate Change (IPCC) and the United Nations Framework Convention on Climate Change (UNFCC) since the early 1990s. The IPCC, he argues, has been "conservative, careful to follow core science rather than policy or advocacy trends" (p. 9). While this obviously remains one of its major strengths, much local evidence of climate change impacts and experience in adaptation […] is gained by local actors and held by civil society actors or published [as gray literature] nationally or regionally and so has been difficult to include. (p. 9)

Local governments like Seattle are, in theory, well positioned to integrate "local evidence" of climate change impacts and experience from an active political ecosystem of nonprofits, neighborhood-based groups, and various environmental and social justice advocacy organizations (such as the "Climate Challenges Atlas" discussed earlier).

But localization is not free of its own power dynamics. The OSE, as discussed in Chapter 3, has actively advanced stronger links between racial and spatial equity concerns and the urban environment, and particularly new analyses of "who is and is not benefiting from Seattle's environmental progress" (see https://www.seattle.gov/environment/). Climate governance and policy work have also moved away from a strong "neighborhood council" system shaped by "white baby boomers who have long dominated discussions about Seattle's future" (Barnett, 2017). In 2016, Mayor Ed Murray directed the Department of Neighborhoods to create a "Community Involvement Commission" and a "Seattle's Renters Commission." Both are designed to help cogenerate new kinds of "risk" information from "under-represented and under-served communities, including, but not limited to, renters, immigrants and refugees, communities of color, people experiencing homelessness, LGBTQ, low-income households, youth and seniors" (Office of the Mayor, 2016).

These formal institutional reforms, championed by the mayor's office, matter for the equitable cogeneration of risk information, but recent developments in "smart city" infrastructures and capabilities will probably matter even more in the coming years. As first discussed in Chapter 2, notions of the smart city refer to how communities like Seattle will increasingly deploy digital, communication, and information technologies (ICT) to occasion spaces of "ubiquitous engagement" (Roy, 2017)—people-to-people; people-to-things; things-to-things; and, possibly, places-to-places. Previously unimagined forms of information, therefore, will be generated and selectively harvested, with still unknown benefits and costs for climate change mitigation and adaptation (Deng et al., 2017).

For some, however, smart city interventions tend to reproduce "corporate storytelling" if not actively democratized (Söderström et al., 2014)—a problem that Haarstad (2017) sees as romanticizing the technocratic conception of urban sustainability and social change "with very limited involvement of societal actors" (cf. Allwinkle & Cruickshank, 2011; Hollands, 2008, 2015). Although the urban planning profession—like other fields—first experienced the technological allure of computational powers back in the 1960s, the veritable explosion of ICT capabilities in the 2000s, and especially the globalized deployment of smartphones between 2001 and 2007, eventually launched dozens of digital services into urban management and policy design: smart apps, automated grids, real-time information flows, embedded sensors, interactive dashboards, Big Data, and so on.

The Milanese-based researchers Corinna Morandi, Andrea Rolando, and Stefano di Vita (2016), argue that the unrealized benefits of novel digital services for improved environmental management and social cohesion could outweigh legitimate concerns with "corporate storytelling" (ibid.) *if* smart cities are reimagined as smart regions; *if* the Internet of Things transforms into an internet of places; and *if* key urban nodes—transit hubs, urban villages, university campus, and so on—are redeveloped into urban digital nodes. Moreover, they see "the Seattle city-region" as one of the most promising places in the United States for such a critical shift:

> Seattle is one of the few and most innovative US smart cities. [...] Despite the weakness of the Seattle city-region in promoting real smart growth (based on the integration of land use and transport through the development of transport nodes) compared with other recent strategic planning visions worldwide, and although this smart city-regionalism based on relations between mobility and the physical space does not seem to exploit the most advanced potentialities offered by ICT's, from a methodological point of view the regional perspective [in the Seattle area] [...] makes it an interesting strategy, which may be able to stimulate a wide-area rescaling of urban smartness. (p. 90)

Such a "rescaling" means further advancing climate change networks—smart or otherwise—and especially strengthening institutions of regional governance and management in the coming years (Dierwechter, 2018).

Advancing networks

Seattle has paid extensive attention for many years to building and/or strengthening "extra-local" networks of climate change governance and policy advocacy. Chapter 3 discussed in some detail Seattle's role in cocreating the national-scale Mayors Climate Protection Agreement under Mayor Greg Nickels as well as Seattle's recent work with the C40 Cities group, arguably the world's most influential transnational municipal network in global-scale carbon diplomacy (Davidson & Gleeson, 2015). Locally built carbon action networks like the King County–Cities Climate Collaboration (K4C) provide further evidence of "multijurisdictional" problem-solving around the search for data, ideas, and improved funding. In addition, as discussed earlier in this chapter, individual municipal departments with green city-regional footprints, notably SPU, participate in various science networks at various scales.

Much more work is needed, however, on how these networks can most usefully shape local policy design and regional program development. Much more work is needed specifically on how initiatives conceived in one place translate into different places, both globally and locally. The Carbon Neutral Cities Alliance (CNCA), for example, is a new collaboration of leading global cities working to achieve carbon neutrality by 2050 or sooner. A recent CNCA project has focused on construction carbon management, aiming to deliver "new policy tools for cities to meet and exceed the World Green Building Council's embodied carbon reduction goals." Seattle is working on this project with Copenhagen, Helsinki, Oslo, San Francisco, Chicago, Toronto, and Vancouver (Carbon Neutral Cities Alliance, 2019).

As indicated in Figure 4.3, such networks can help to integrate global mitigation goals with local adaptation policies. But much more work is needed on how these policy tools spill into and inform regional policy and advocacy networks like K4C and, in my view, more traditional public authorities like the Puget Sound Regional Council or Sound Transit, both of whom are actively involved in the growth management and planning regulation of construction activity in the metropolitan area as a whole. For that to happen, Seattle in turn needs to regionalize even further its vision of local action, collaborating better with nearby cities with whom it also competes for economic investment, a process of building new "epistemic communities" that Chris Benner and Manuel Pastor (2012) see as crucial to advancing a more inclusive prosperity in city-regions like Seattle.

Focusing on disadvantaged populations

The animating heart of a recent statement on improving climate change planning—*Carbon Neutral, Climate Ready* (Office of Sustainability and Environment, 2017)—is framed around deep-seated concerns with "Equity and Climate Change." Previous chapters, as well as the discussion provided here, have covered this theme extensively. In 2017, Seattle and San Francisco shared the unwanted title of "the most unequal city in America," as measured by local Gini coefficients (Balk, 2017).

While Seattle's recent struggles with homelessness and defenseless encampments under a carbon-belching interstate system provide the most dramatic evidence of uneven, unhealthy, and unsustainable development, Seattle also shares the much larger crisis of growing income polarization across most large metropolitan regions of the United States (Himes, 2019). Attacking these complex and long-running structural problems in high-tech regions of the American economy through green policy design, or even through local institutional reforms, has profound limits, particularly without stronger regional political movements and new kinds of federal financial support (Herrschel & Dierwechter, 2018).

Still, Seattle's comparatively progressive urban political culture aptly foregrounds how the city is now attempting to confront its unwanted status as an "elite emerald" rather than "emerald city," how it seeks to become a green city with blue collars rather than a city retrofitted only for a comfortable cognitariat of information age firms (Holmes, 2020). Along with leveraging co-benefits and tapping more into underutilized ecological services, "equity" is the most important "planning priority" for future climate preparedness. The city therefore strategically aims to

> prioritize actions that reduce risk and enhance resilience in frontline communities (e.g., communities of color, lower income communities, immigrant and refugee communities, disabled residents and seniors), as they are at greater risk from the impacts of climate change and often have the fewest resources to respond to changing conditions. (Office of Sustainability and Environment, 2017, p. 3)

That is because, as one example,

> high temperatures on [wealthy] Queen Anne hill do not come with the same implications that accompany similar temperatures in the [more vulnerable] International District where air pollution levels increase the likelihood of negative impacts. Likewise, a flood in the lower Duwamish area where residents may have fewer resources to repair damages to their homes will affect residents differently than some higher income residents in West Seattle. (p. 16)

What is "needed" in Seattle to close the yawning gaps between extant problems and partial solutions? According to the city, a thousand small-bore, ordinary, generally unglamorous governance efforts, such as:

- "Broader conversations" with community-based organizations (p. 3).
- "Additional research" on the precise downscaling impacts of rising sea levels, extreme wind events, and other hazards (p. 10).
- "Changes" to public facility design standards given increasing temperature and precipitation regimes (p. 24).

- "Better evaluation" of resource supply and demand in electricity grids for the next 20 years through more precise models and smart data flows (p. 42).
- "Developing a funding strategy" for a comprehensive public health and climate change program to include data monitoring and surveillance systems to track and report human health effects related to climate change, particularly for disproportionally impacted population (p. 64).
- "Providing support for safety-net programs" that improve access to healthy food for low-income residents in the case of rising food prices (p. 66).
- "Fostering social cohesion" through daily neighborliness to promote pragmatic resilience in the face of extreme weather and other climate challenges (p. 76).

All this and much more—on other lists, in other plans, in and beyond Seattle—is surely needed. But is this what is *most* needed? Armed at a safe distance with Foucault, Marx, or Weber, critical urban scholars of climate action—including me—can easily dismiss this prosaic, frenetic, quotidian activity as little more than fiddling while Rome burns. Perhaps critical urban scholars are right, scraping away superficial veneers of the "real world" to reveal far more compelling social truths and institutional insights. The most frightening of these may be that, as one scholar plainly argues, it is "too late for sustainability" (Bullard, 2011). But putting out small fires, one at a time, surely can make a difference if enough people are putting out small fires. What is most needed, then, is not only a new set of carbon policies but a new politics of urban citizenship (Karvonen, 2011).

Conclusion

The globe's climate has changed many times in the Earth's long history. At different points in time, mean temperatures have been both higher and lower than at present, oscillating "naturally" to complex factors or responding violently to singular events, like the Chicxulub asteroid strike that killed off the dinosaurs (and paradoxically favored the evolution of mammals). Regions have changed as well; the Sahara Desert used to grow wild grains. Contemporary climate change, however, is not natural. Nor is the possibility of "self-extinction" in a few generations and perhaps much earlier (Kolbert, 2014). For contemporary climate change is a "hybrid condition" of the Anthropocene, in general, and the relatively recent carbonization of urban capitalism, in particular.

"Decarbonizing" late capitalist cities like Seattle—building post-carbon socio-technical systems that structure daily life—in order to mitigate the global severity of climate changes and to adapt locally to these same changes requires what the ARC3.2 framework conceptualizes (if not quite theorizes) as an interrelated set of urban transformations. On its own and especially in novel networked partnerships with community groups, other cities, states, and international organizations, Seattle arguably has a fighting chance to soften the treacherous triptych of increasing temperatures, more intense precipitation patterns, and rising sea levels. Failure to do so, though, will vastly amplify the city's many existing social vulnerabilities and disproportionately impact its most disadvantaged citizens.

Mitigation concerns in and around Seattle are now almost 30 years old, and more recent efforts to integrate mitigation with adaptation policies will grow more prominent. Efforts to merge DRR with climate change policies are, in my view, less developed but could improve quickly if, for example, local smart growth plans in Seattle and adjacent communities steadily evolve into more creative place-making forms associated with what Honachefsky (2000, p. 13) calls "ecologically-based municipal land-use planning." As he suggests, this new form of planning "can be accommodated within the existing land use tools and practices already in place." Although the mitigation effects of smart growth plans have received some attention in both the gray and scholarly literatures, "very little has been done to understand the association between climate change, smart growth policies, and disaster risk reduction strategies at the local level" (Chatterjee et al., 2020, p. 68). Along with metropolitan Portland, Greater Seattle nonetheless represents one of the most important places to consider these transitions in the coming years. Innovative projects at scale are already emerging across the region (Vincent, 2019).

Activating such a transformation will require new ways of generating risk information, not only for human systems but from the wider socio-ecological assemblages within which these systems exist. Citizen scientists can help, but a more generalized democratization (and bioregionalization) of "smart city" technologies can help even more. "Real-time" information around the benefits, costs, and temporalities of transit use, electricity consumption, waste production, raingarden services, and so on represent new opportunities for political communities to harness together science and citizenship in the service of urgent carbon action.

Seattle's recent political embrace of a "local Green New Deal," first discussed in Chapter 3, provides recent evidence of the city's long-term potential to occasion such shifts. So does the older history of SPU in addressing concerns like winter storm surges and summer droughts. The manifold advantages afforded since 2004 by a carbon-neutral energy provider like Seattle City Light are also critical to consider.

Can these efforts advance the equity of climate change planning in the coming years and decades? The city is certainly trying to find out. At a 2017 meeting of Seattle's Community Advisory Technology Committee, volunteer members received an update on the city's new Opportunity Zone Prioritization Project:

> Over the summer, Faisal Jama from East African Community Services (EACS) will create a capacity building playbook mapping out non-profit organizations working with underserved communities in Seattle's Southwest region (the 98118 zip code). Mr. Jama hopes to build a tool that can be used by companies seeking to partner with nonprofits in this region, and replicated throughout the region by other communities. (Community Technology Advisory Board, 2017)

This work highlights the kinds of future state–civil society relationships most needed to democratize urban change and thus widen both information and value fields about justly managing global climate change. Mr. Jama's efforts suggest the unrealized potential of "community-based regionalism," as grassroots actors in one community aim to link up with grassroots actors in other communities, creating a social counterweight to business leaders, who mainly see the city-region as the level at which firms cluster to compete, and environmental advocates, who suggest that only region-wide planning can stem sprawl and save open space (Pastor et al., 2004, p. 4). Coupled with "top-down" mayoral and council leadership and "meso-level" professional capacity, it indicates how an urban transformational agenda for climate action can address Seattle's social vulnerabilities through inclusive outreach.

Chapter 5

CONCLUSION: SEATTLE'S LESSONS

> Real generosity towards the future lies in giving all to the present.
> Albert Camus (1963)

Cities are social, cultural, and political spaces forged from the economic conversion of nature's bounty into specific forms of human development. Differently situated groups of people differently experience these forms. However talented, individuals are refracted through race, class, gender, and other social frames of identity and (dis)empowerment. Yet all are facing the calamitous reality of global climate change. As the ARC3.2 report of the Urban Climate Change Research Network specifically puts it, cities are, in their challenging phrase, "complex social-ecological systems" (UCCRN, 2018, p. 7). Innovative and effective ideas for how political and policy communities at various scales with diverse interests and values can manage these systems relate to at least five areas of particular concern: integrating carbon mitigation with adaption; coordinating disaster risks with climate adaptation; cogenerating and deploying risk information more effectively; advancing trans-local action networks; and, not least, focusing on disadvantaged populations who are particularly vulnerable.

Cities like Seattle are legally bounded places with clear jurisdictions, yet also sinuous, "stringy," web-like relationships and bundled connections, such as city-regional economies of carbon-using commuters and trans-local ecologies of water flows and energy grids that "stretch" Seattle's urban experience well beyond these legally bounded places (see Figure 1.2). So, too, are the many transnational municipal networks within which Seattle now actively participates, such as C40 Cities and, more recently, the Carbon Neutral Cities Alliance, that each seek collective urban action at the global scale. Seattle has not been a passive bystander but a dynamic cocreator of these new spaces.

After Donald Trump withdrew the United States from the Paris Agreement in early June 2017, Seattle immediately passed a remarkable resolution to uphold its financial portion of America's national commitments. Seattle directed municipal support for the Green Climate Fund (GCF). "The Seattle resolution appears to be unique in its support for the fund," Karl Mathiesen (2017) observed soon thereafter:

> Hundreds of state and local government entities have come forward in recent weeks to restate their own commitment to the Paris deal, including some that have passed official laws or statements. But most have focused on emissions reductions, rather than the obligations of rich countries to the world's poor. [...] The [Seattle] council resolution said it would work with concerned communities, companies and local and state governments to fulfil the commitments made by the US government. The only other non-national governments to have contributed to the GCF are three regions of Belgium, [...] Flanders, Wallonia and the city region of Brussels.

Working to cocreate "post-carbon" cities financially (and geopolitically) is thus a never-ending series of interrelated projects that challenge traditional ways of thinking about how "cities" can and do contribute to what now arguably constitutes the gravest threat to human life on Earth as presently understood (Taylor et al., 2020).

"Traditional" approaches still matter, too. In 2014, Seattle voters overwhelmingly approved a proposition to fund improved transit programs through the Seattle Transportation Benefit District (STBD). New funds from car license fees and local sales taxes enhanced transit trips and service hours on weekdays, nights, and weekends and significantly expanded the percentage of households within a 10-minute walk of transit services. In addition, the STBD provided improved bus access to public high school students and helped the Seattle Housing Authority to distribute free passes to low-income residents while supporting community-based programs to increase mobility for seniors and people with disabilities. The effects on mode share to the downtown area of Seattle were especially dramatic. Single-occupancy vehicle use dropped from 50 percent to 25 percent between 2000 and 2017, while bus use rose from 20 percent to 49 percent (Fesler, 2019).

No doubt, then, there is something to admire in Seattle's record on sustainability issues in general and global climate change in particular. In part this is a question of "luck," as Guy Lawrence (1996, p. 111) has noted, but "good decisions" are vital. Over 90 percent of Seattle's electricity is generated by (distant) hydroelectric dams, yet it is important that in 2005 Seattle City Light became the first electric utility in the United States to achieve zero net greenhouse gas emissions. In the planning arena, concerns with more sustainable forms of urban development in the 1990s, which built on local conservation efforts dating back to the 1950s, actively shaped Seattle's landmark comprehensive plan in 1994. That plan promulgated long-range efforts to manage high-tech growth around "urban villages" and "centers" and inaugurated the (then novel) policy language around "climate-altering greenhouse gases" and "carbon-neutral" development patterns. King County later pioneered the regulatory use of greenhouse gas auditing in permit reviews and furthermore worked with the City of Seattle, the Port of Seattle, and numerous other jurisdictions to coordinate local climate action across the metropolitan area, including new efforts to lobby for more state funds and green investments. In 2000, Seattle became the first city in the United States to establish green building codes for municipal facilities; a year later, the city created a Leadership in Energy and Environmental Design (LEED) program for private development (City of Seattle, 2013).

Strong mayoral leadership over the past several years has consistently championed local support for global climate action, notably under Greg Nickels, and city council leadership has actively supported local administrative reforms, such as creating the Office of Sustainability and Environment (OSE) under Mayor Paul Shell in 2000; this has culminated most recently in clear urban policy commitments to pursuing, as in Los Angeles and a few other cities, a local "Green New Deal" that remains politically difficult at the federal level within the United States. Although only a tiny part of the city's budget, the OSE has helped to develop, coordinate, and monitor interdepartmental climate change—providing institutional focus and normative clarity. It has spearheaded mitigation policies and adaptation programs that have grown increasingly more prominent in recent years. One example is the Department of Finance and Administrative Services' (2019) recent efforts at municipal fleet electrification and improved use of fossil-free fuels.

Progressive local politics in Seattle reflect a vigorous and well-organized social ecosystem of highly skilled nonprofits, community-based organizations, activist groups, and neighborhood associations, particularly those that interpret environmental concerns through social justice goals. Since 1991, for instance, the nonprofit organization Sustainable Seattle has pioneered the use of region-wide indicators for measuring what "sustainability" might actually mean on the ground, influencing local approaches in Seattle, above all, but also green assessment metrics in cities around the world. Capitol Hill Housing, a community development corporation, has worked with Seattle City Light to advance the Capitol Hill Eco-District Project, which also includes efforts to support a new sharing economy (making it easier to understand why the anarcho-radical Capitol Hill Organized Protest (CHOP) momentarily emerged in Seattle in the tragic wake of the George Floyd protests). Still other nonprofit groups, such as Puget Sound Sage and GotGreen!, have pushed for more economically equitable and racially inclusive forms of transit-oriented development and related urban changes, especially in strategic areas of the city undergoing acute gentrification and displacement pressures. As one volunteer with GotGreen! put it in ways redolent of Seattle's wider activist and at times even messianic political culture:

> If we take climate change seriously, we have an opportunity to not only save ourselves from the biggest threat humanity has faced collectively [...] but we also have the opportunity to right the wrongs of capitalism, white supremacy, and patriarchy. (see https://gotgreenseattle.org/home/what-we-do/#climatejustice)

In some visions of Seattle, climate responses are what Harriet Bulkeley (2013) calls "evolving resilience," produced from the local activities and self-organization of independent, decentralized, often autonomous actors who are focused on "radical possibilities for living different urban lives in reconfigured urban economies" (p. 11). These "bottom-up" visions of future, livable cities—one set of approaches to Seattle's future—contrast in theory with what Bulkeley sees as more "top-down" visions of "climate control" achieved through "combinations of managed change and technical innovation"; here the key actors are various state institutions and, it seems, eco-innovative companies (p. 14).

In practice, as Seattle's empirical experience with climate action shows, partnerships, associations, and collaborations—though never easy, smooth, or fully realized—suggest that dichotomous, either/or readings of "bottom-up" activism versus "top-down" systems capture rather little of what will matter on the ground over time: namely, inter-scalar, multi-sectoral, multi-actor coalitions that actually remake specific aspects of the larger "socio-technical systems" shaping (re)urbanization patterns across "vulnerable" bioregions (Georgiadis, 2020). Seattle can learn here from research produced over many years by the Baltimore School of Urban Ecology, for example, including green governance insights into how "stewardships networks" between state and society emerge and become efficacious across wider watersheds as well as how new ideas for urban ecological design and smart regulatory land-use strategies, including sprawl containment, intersect with social and ecosystem dynamics (Pickett et al., 2019). These include the legacy effects of past decisions, the unintended outcomes of density policies on urban runoff, tensions between municipal autonomy and regional interdependencies, and the segmentation of data (Irwin et al., 2019).

Large-scale, durable shifts in political development further stand out. As discussed in Chapters 2 and 3, state-legislative reforms, such as Washington's

Figure 5.1. Integration of water management with urban planning and urban design in suburban Seattle (Source: photo by author, 2019)

Growth Management Act in 1990/91, have created legal planning frameworks for core cities like Seattle to reshape modernist space economies and (albeit slowly) link with city-regional transit agencies, such as Sound Transit, along with metropolitan planning organizations (MPOs) like the Puget Sound Regional Council (PSRC). Most MPOs in the United States have had very little to say so far about climate mitigation and adaptation or urban resiliency policies (Mason & Fragkias, 2018). As the PSRC, which is based in downtown Seattle, now moves more directly into global climate action, new opportunities for ecologically based planning and climate-friendly transit in diverse communities will likely emerge, especially around what Raven et al. (2018, p. 155) call the "integration of water management with urban planning and urban design" (Figure 5.1).

Yet tensions pervade this process, and anxieties over limited results run high. Seattle has put "activists in city hall" (Clavel, 2010) at crucial times in recent decades, even as world-famous billionaires from Microsoft cofounder Paul Allen to Amazon's Jeff Bezos have accelerated the "radical reurbanization" of the urban core through high-tech capitalism (Gregory, 2015)—elevating the role of the "cognitariat" in the social geographies of the city (Scott, 2011; Gibson, 2004). As discussed in Chapter 3, Seattle has nurtured new forms of "climate disobedience" (Burkett, 2016), such as the "kayaktivists" protests against Shell oil and Arctic drilling.

Long shaped by the manufacturing prowess of Boeing, which makes products long associated with rising carbon emissions (Åkerman, 2005), the Seattle–Tacoma International Airport nonetheless has been a leader since 2011 in the search for, and accelerated uses of, aviation biofuels (Stanton, 2017). The Port of Seattle has worked with Governor Jay Inslee—America's most vocal "Climate Governor"—and many other actors to try to create "a new network of interlocking infrastructures [...] connecting farmers with refiners, distributors and users" (Boyle, 2019). The Port of Seattle is specifically positioning itself to play a leading role in leveraging new "funding models" for aviation biofuels that substantially cut carbon emissions, including corporate support, port taxing authority, the uses of general non-aeronautical revenue, and airline agreements (Klauber et al., 2017, pp. 18–20). All hugely difficult, such projects are perfect examples of the "combinations of managed change and technical innovation" that, again, Bulkeley associates with more "top-down" approaches to urban climate governance.

For radical actors in Seattle more comfortable imagining alternative economies in community halls than ecological modernization in corporate boardrooms, such developments fall well short of the hypothesized "urban transformation" suggested in the ARC3.2 report (Figure 1.1). Greening aviation fuels mitigates global carbon through the largely unchanged behaviors of privileged travelers but does little to envision what McLaren and Agyeman (2018) call the "sharing city," an urban space economy and "solidarity society" reshaped by consumer and housing cooperatives, community currencies, democratic employee stock ownership, shared land ownership, and cooperative banks, among others.

The recent rise of "coworking" facilities reflects these tensions. Located in Seattle's challenged Rainier Valley, for instance, "The Hillman City Collaboratory" is an incubator that caters to nonprofit organizations, startup movements, and otherwise isolated individuals with a social change mission. Here coworking provides an inspirational, inclusive, affordable space where many groups and individuals "can build community and work for social change" (Hillman City Collaboratory, 2019). In consequence, "innovation" is less economic than political, and "development" is about social justice and green growth. But other coworking spaces are franchised offices of national corporations like WeWork, headquartered in New York City, or Regus, headquartered in Brussels. These locations reinforce the architectonics of car-centered office complexes, contributing less to socially embedded place-making or community-building dynamics associated with "sharing cities" that represent just transitions to climate-friendly urbanism.

Hence, Seattle's multiple lessons: some focused on progressive-liberal idea(l)s associated with "carbon control," often techno-managerial in orientation; others, on "evolving resiliency" through more critical political engagements with economy and society. Innovation happens in both arenas, of course, and "overlaps" between them are emergent, as evidenced empirically in stormwater management projects like Seattle Public Utilities' Street Edge Alternatives ("SEA Streets"), which has inaugurated new forms of natural drainage systems in residential neighborhoods and, hence, new forms of "shared" green commons (Liptan, 2017, pp. 58–60).

Taken in aggregate, as first discussed in Chapter 2, by 2016 Seattle had reduced local emissions by 5 percent in toto even as the city's population grew 18 percent—a decrease of 20 percent in per-person emissions. Although these figures exclude emissions from the original production of Seattle's consumed goods and services, the municipality champions itself as "one of the most climate friendly cities in the nation" (City of Seattle, 2016c, p. 2). In the transport sector, for example, road emissions per capita (in millions of metric tons of CO_2e) declined 18 percent between 1990 and 2018; per capita emissions in residential and commercial buildings for the same period declined 30 percent (ibid., see table 20, p. 36). In late 2014, based on these data, the Obama administration considered Seattle "on the frontier of ambitious climate action," whose approaches "provide a model for other communities to follow" (White House, 2014).

Yet the Seattle city council roundly judged its own efforts "insufficient," particularly vis-à-vis the larger goal of making Seattle's economy "more equitable and ecologically sustainable" (ibid.). Moreover, as elaborated in Chapter 4, current climate models of temperature changes, projected sea-level rise, and precipitation patterns for the Seattle area suggest dramatic alterations in the coming decades, even under the most optimistic assumptions about overall global mitigation policies. While mitigation efforts remain crucial (City of Seattle, 2013), policies in Seattle are now transitioning rapidly to integrated adaptation efforts (City of Seattle, 2017a), albeit unevenly across departments (Pacino, 2014) to say nothing of the wider city-region, where socio-technical changes remain daunting (Abel & White, 2015; Dierwechter, 2010, 2017).

Such efforts increasingly demand, as already apparent in the nascent and intriguing K4C experience discussed in Chapter 4, the turn away from "methodological cityism" (Angelo & Wachsmuth, 2014) and toward more advanced and effective forms of green metropolitan regionalism (Herrschel & Dierwechter, 2018). Air, water, energy, stormwater, housing, transit, and certainly disasters—all these flow across municipal borders (Figure 1.2). Borrowing a phrase from Mossner and Miller (2015), cities like Seattle are not actors in charge of their own sustainability; they are places whose successes are related to the role(s) of others in dense city-regional complexes that draw on bioregional assets for development. In my judgment, "urban" climate action must regionalize much further than it has for transformation to progress, as "regional decisions may affect neighborhoods or individual parcels and vice versa" (Raven et al., 2018, p. 140).

In part, transformation will come "from below," originating with the community-based regionalism of activists linking their work across borders and watersheds; in part, it will come "from above," as state legislatures and federal authorities create novel policy environments that encourage more effective cooperation around local climate action, empowering metropolitan planning organizations, for instance, that link transit and land-use policies to carbon mitigation and adaptation efforts (Bakkenta, 2020). Transformation will also come from the "sideways" politics (or urban geopolitics) of cities-in-networks, teaching and learning and ultimately challenging long-standing forms of authority (Dierwechter, 2019, chapter 3).

Following Albert Camus, generous efforts in "giving all to the present" are undoubtedly required to secure a future worth having for ourselves and especially for youth and those yet to come. In Seattle, as I have argued throughout this book, that future must increasingly focus on ameliorating a deeply troubling contradiction. Seattle in the 2020s is less an "Emerald City" than an "Elite Emerald." Income inequalities have grown; gentrification pressures have increased; class structures have steadily shifted upward, leaving the working poor and homeless especially vulnerable to climate change. Seattle cannot become post-carbon if it is not also post-polarized; resilient if not also just. Seattle must become, in a word, "Everyone's Emerald," a city that also seeks working-class sustainability by embracing its progressive reputation and amplifying the "loud echoes of a more radical past" (Gregory, 2015). The lessons that Seattle learns in pursuit of more inclusive climate action will thus be of abiding interest to cities and metropolitan regions across the United States and all around the world.

REFERENCES

350 Seattle. (2018). *350 Seattle 2018 Annual Report.* Seattle: 350 Seattle.
Abbott, C. (1992). Regional city and network city: Portland and Seattle in the twentieth century. *Western Historical Quarterly, 23*(3), 293–322. doi:10.2307/971508.
Abel, T., & White, J. (2015). Gentrified sustainability: Inequitable development and Seattle's skewed riskscape. *Interdisciplinary Environmental Review, 16,* 124–57.
Abel, T., White, J., & Clauson, S. (2015). Risky business: Sustainability and industrial land use across Seattle's gentrifying riskscape. *Sustainability, 7*(11), 15718–53.
Acuto, M. (2013). The new climate leaders? *Review of International Studies, 39*(4), 835–57. doi:10.1017/s0260210512000502.
Angelo, H., & Wachsmuth, D. (2014). Urbanizing urban political ecology: A critique of methodological cityism. *International Journal of Urban and Regional Research, 39*(1), 16–27.
Åkerman, J. (2005). Sustainable air transport—on track in 2050. *Transportation Research Part D: Transport and Environment, 10*(2), 111–26. doi:10.1016/j.trd.2004.11.001.
AlAwadhi, S., & Scholl, H. J. (2013, January 7–10). Aspirations and realizations: The smart city of Seattle. Paper presented at the 2013 46th Hawaii International Conference on System Sciences, Wailea, Maui, HI, USA.
Alkon, A. H., & Mares, T. M. (2012). Food sovereignty in US food movements: Radical visions and neoliberal constraints. *Agriculture and Human Values, 29*(3), 347–59. doi:10.1007/s10460-012-9356-z.
Allwinkle, S., & Cruickshank, P. (2011). Creating smart-er cities: An overview. *Journal of Urban Technology, 18*(2), 1–16. doi:10.1080/10630732.2011.601103.
Altshuler, A. (1965). The goals of comprehensive planning. *Journal of the American Institute of Planners, 31*(2), 186–95.
Andrews, K. T., & Edwards, B. (2004). Advocacy organizations in the U.S. political process. *Annual Review of Sociology, 30*(1), 479–506. doi:10.1146/annurev.soc.30.012703.110542.
Atkinson, S. (2017). Public lecture. City of Tacoma, UW Tacoma, April 26. Baker, J. (2020). Center for Urban Waters & Puget Sound Institute, Tacoma. Personal communication, March 3.
Bakkenta, B. (2020). Puget Sound Regional Council. Zoom interview (conducted by Doug Carlson and Chris Moradi). April 15.
Bakkente, B., Lawlor, M. P., Roberts, A., & Redinger, M. (2015). Growing transit communities. In J. Sterrett, C. Ozawa, D. Ryan, E. Seltzer, & J. Whittington (eds.), *Planning the Pacific Northwest.* Chicago: APA Press.
Balk, G. (2014, November 12). As Seattle gets richer, the city's black households get poorer. *Seattle Times.* Retrieved from http://blogs.seattletimes.com/fyi-guy/2014/11/12/as-seattle-gets-richer-the-citys-black-households-get-poorer/.

———. (2017, November 17). Seattle hits record high for income inequality, now rivals San Francisco. *Seattle Times*. Retrieved from https://www.seattletimes.com/seattle-news/data/seattle-hits-record-high-for-income-inequality-now-rivals-san-francisco/.

———. (2019, November 2). Seattle's rate of car ownership saw the biggest drop among big U.S. cities—by far. *Seattle Times*. Retrieved from https://www.seattletimes.com/seattle-news/data/seattles-rate-of-car-ownership-saw-the-biggest-drop-among-big-u-s-cities-by-far/.

Barber, B. (2013). *If Mayors Ruled the World: Dysfunctional Nations, Rising Cities*. New Haven, CT: Yale University Press.

Barbour, E., & Deakin, E. A. (2012). Smart growth planning for climate protection evaluating California's Senate Bill 375. *Journal of the American Planning Association*, *78*(1), 70–86. doi:10.1080/01944363.2011.645272.

Barnett, E. (2017, April 3). How Seattle is dismantling a NIMBY power structure. *Next City*. Retrieved from https://nextcity.org/features/view/seattle-nimbys-neighborhood-planning-decisions.

Barrutia, J., & Echebarria, C. (2011). Explaining and measuring the embrace of Local Agenda 21s by local governments. *Environment and Planning A*, *43*(2), 451–69. doi:10.1068/a43338.

Bassett, E., & Shandas, V. (2010). Innovation and climate action planning. *Journal of the American Planning Association*, *76*(4), 35–450.

Beauregard, R. (2018). *Cities in the Urban Age: A Dissent*. Chicago: University of Chicago Press.

Beauregard, R. A. (2012). Planning with things. *Journal of Planning Education and Research*, *32*(2), 182–90. doi:10.1177/0739456x11435415.

Beekman, D. (2019a, March 18). Seattle upzones 27 neighborhood hubs, passes affordable-housing requirements. *Seattle Times*. Retrieved from https://www.seattletimes.com/seattle-news/politics/seattle-upzones-27-neighborhood-hubs-passes-affordable-housing-requirements/.

———. (2019b, November 9). Sawant surges past Orion in Seattle City Council race with Friday vote counts. *Seattle Times*. Retrieved from https://www.seattletimes.com/seattle-news/politics/sawant-takes-lead-over-orion-in-seattle-city-council-race-with-friday-vote-count/?amp=1.

Bell, K. (2019, October 11). A working-class green movement is out there but not getting the credit it deserves. *Guardian*. Retrieved from https://www.theguardian.com/environment/2019/oct/11/a-working-class-green-movement-is-out-there-but-not-getting-the-credit-it-deserves.

Benner, C., & Pastor, M. (2012). *Just Growth: Inclusion and Prosperity in America's Metropolitan Regions*. London: Routledge.

Blair, J. K. (2014). *Fire and Gold Build Seattle* (undergraduate thesis), University of Washington Tacoma, Tacoma.

Blue Ribbon Commision on Sustainability. (2009). *Greening Mass Transit and the Metro Regions*. New York: New York Metropolitan Transit Authority.

Bouteligier, S. (2012). *Global Cities and Networks for Global Environmental Governance*. Hoboken, NJ: Taylor and Francis.

Boyle, A. (2019, March 6). Port of Seattle pushes for sustainable aviation fuels, but it's not easy being green. Retrieved from https://www.geekwire.com/2019/port-seattle-pushes-sustainable-aviation-fuels-not-easy-green/.

Brand, P. C. (2005). *Urban Environmentalism: Global Change and the Mediation of Local Conflict*. London: Routledge.

Brown, F. L. (2016). *The City Is More Than Human: An Animal History of Seattle*. Seattle: University of Washington Press.

Browning, C. R., Wallace, D., Feinberg, S. L., & Cagney, K. A. (2006). Neighborhood social processes, physical conditions, and disaster-related mortality: The case of the 1995 Chicago heat wave. *American Sociological Review, 71*(4), 661–78. doi:10.1177/0003122406071000407.

Brunner, R. (1991). Global climate change: Defining the policy problem. *Policy Sciences, 24*(3), 291–311.

Bulkeley, H. (2013). *Cities and Climate Change*. London: Routledge.

Bulkeley, H., & Betsill, M. M. (2005). Rethinking sustainable cities: Multilevel governance and the "urban" politics of climate change. *Environmental Politics, 14*(1), 42–63. doi:10.1080/0964401042000310178.

Bullard, N. (2011). It's too late for sustainability. What we need is system change. *Development, 54*(2), 141–42.

Burkett, M. (2016). Climate disobediance. *Duke Environmental Law & Policy Forum, 27*(1), 1–50.

Burnham, M., Ma, Z., & Zhang, B. (2016). Making sense of climate change: Hybrid epistemologies, socio-natural assemblages and smallholder knowledge. *Area, 48*(1), 18–26. doi:10.1111/area.12150.

C40 Cities. (2019). Seattle Mayor Jenny Durkan elected to the C40 Cities Steering Committee (press release). Retrieved from https://c40-production-images.s3.amazonaws.com/press_releases/images/435_Seattle_SC_Announcement.original.pdf?1575991614.

Callon, M. (1984). Some elements of a sociology of translation: Domestication of the scallops and the fishermen of St Brieuc Bay. *Sociological Review, 32*(Suppl 1), 196–233. doi:10.1111/j.1467-954X.1984.tb00113.x.

Camus, A. (1963). *Notebooks, 1935–1942*. New York: Knopf.

Cannon, L. (1977, May 19). An upset in Seattle; "jobs" wins over "environment" as Republican defeats favorite for Brock Adams' seat. *Washington Post*, p. A2.

Capitol Hill Housing. (2018). *Capitol Hill Eco-District 2018 Community Report*. Retrieved from https://issuu.com/mseiwerath/docs/ecodistrict_2018-final-web.

Carbon Neutral Cities Alliance. (2019). Policy framework to dramatically reduce embodied carbon in cities developed to respond to climate goals (press release). Retrieved from http://carbonneutralcities.org/wp-content/uploads/2019/09/Embodied-Carbon-Policy-Framework-for-Cities-Press-Release_FINAL_26.09.19.pdf.

Causone, F. (2018). Climate change: Unleashing the geopolitical power of cities. Retrieved from https://www.ispionline.it/en/pubblicazione/climate-change-unleashing-geopolitical-power-cities-21556.

Center for Operational Oceanographic Products and Services. (2017). *Global and Regional Sea Level Rise Scenarios for the United States*. Silver Spring, MD: US Department of Commerce, National Oceanic and Atmospheric Administration, National Ocean Service, Center for Operational Oceanographic Products and Services.

Chatterjee, V., Andrew, S., & Feiock, R. (2020). Linking smart growth policies and natural disasters: A study of local governments in Florida. In S. Gerrin & R. Brinkmann (eds.), *Case Studies in Suburban Sustainability*. Gainesville: University of Florida Press.

City of Seattle. (1994). *Toward a Sustainable Seattle: A Plan for Managing Growth, 1994–2014; the City of Seattle Comprehensive Plan*. Seattle: Seattle City Council.

———. (1997). *Ordinance No. 118597*. Seattle: Office of the City Clerk. Retrieved from http://clerk.seattle.gov/search/ordinances/118597.

---. (2006a). *A RESOLUTION Setting Forth the 2006 State Legislative Agenda of the City of Seattle*. Seattle: Office of the City Clerk. Retrieved from http://clerk.seattle.gov/~archives/Resolutions/Resn_30830.pdf.

---. (2006b). *Seattle, a Climate of Change: Meeting the Kyoto Challenge*. Seattle: Office of the Mayor.

---. (2009). *US Mayors Climate Protection Agreement*. Retrieved from http://www.seattle.gov/mayor/climate/quotes.htm.

---. (2010). *Seattle Climate Action Highlights 2010*. Seattle: Office of Sustainability and Environment.

---. (2013a). *Seattle Climate Action Plan*. Retrieved from http://www.seattle.gov/Documents/Departments/Environment/ClimateChange/2013_CAP_20130612.pdf.

---. (2013b). *A RESOLUTION Adopting the 2013 Seattle Climate Action Plan*. Retrieved from http://clerk.seattle.gov/~archives/Resolutions/Resn_31447.pdf.

---. ([1994, 2005] 2015). *Comprehensive Plan: Towards a Sustainable Seattle: A Plan for Managing Growth 2015–2035*. Seattle: Seattle Planning and Development.

---. (2016a). *Equitable Development Implementation Plan*. Retrieved from https://www.seattle.gov/Documents/Departments/OPCD/OngoingInitiatives/SeattlesComprehensivePlan/EDIImpPlan042916final.pdf.

---. (2016b). *Seattle 2035*. Retrieved from https://www.seattle.gov/Documents/Departments/OPCD/OngoingInitiatives/SeattlesComprehensivePlan/SeattleComprehensivePlanCouncilAdopted2016.pdf.

---. (2016c). *Seattle Community Greenhouse Gas Emissions Inventory*. Seattle: Office of Sustainbility and Environment.

---. (2017a). *Carbon Neutral, Climate Ready: Preparing for Climate Change*. Seattle: Office of Sustainability and Environment.

---. (2017b). *Office of Emergency Mangament Annual Report*. Seattle: Office of Energy Management.

---. (2018). *Seattle Climate Action*. Seattle: Office of the Mayor.

---. (2019). *Resolution 31895, V2*. Seattle: Seattle Office of the City Clerk. Retrieved from https://seattle.legistar.com/ViewReport.ashx?M=R&N=Text&GID=393&ID=3611579&GUID=ADF51F71-1823-4D7B-B599-9ED04DFD8860&Title=Legislation+Text.

---. (2020a). *2020 Proposed Budget*. Seattle: City Budget Office. Retrieved from http://www.seattle.gov/city-budget-office/budget-archives/2020-proposed-budget.

---. (2020b). *Seattle Hazard Explorer*. Retrieved from http://seattlecitygis.maps.arcgis.com/apps/MapSeries/index.html?appid=0489a95dad4e42148dbef571076f9b5b.

City of Vancouver. (2013). *Vancouver's Digital Strategy*. Retrieved https://vancouver.ca/files/cov/City_of_Vancouver_Digital_Strategy.pdf.

Claire, C. (2013). Framing the end of the species. *Symplokē*, 21(1–2), 51–63. doi:10.5250/symploke.21.1-2.0051.

Clapp, J., & Dauvergne, P. (2008). *Paths to a Green World: The Political Economy of the Global Environment*. Cambridge, MA: MIT Press.

Clark, A. (2019). Director of Strategic Partnerships, Stewardship Partners. Personal communication.

Clavel, P. (2010). *Activists in City Hall: The Progressive Response to the Reagan Era in Boston and Chicago*. Ithaca, NY: Cornell University Press.

Clear Capital. (2019). *Home Data IndexTM (HDI)*. Retrieved from https://www.clearcapital.com/wp-content/uploads/2017/09/Home-Data-Index.pdf.

Cleveland, J., & Plastrik, P. (2018). *Life after Carbon: The Next Global Transformation of Cities*. Washington, DC: Island Press.

Coenen, L., Benneworth, P., & Truffer, B. (2012). Toward a spatial perspective on sustainability transitions. *Research Policy, 41*(6), 968–79. doi:https://doi.org/10.1016/j.respol.2012.02.014.

Coletta, C., Evans, L., Heaphy, L., & Kitchin, R. (2019). *Creating Smart Cities*. Abingdon, Oxon: Routledge.

Community Technology Advisory Board. (2017). *[CTAB-EGOV] eGovernment Committee Meeting Minutes—5/30/2017*. Seattle: City of Seattle. Retrieved from https://ctab.seattle.gov/2017/05/31/ctab-egov-egovernment-committee-meeting-minutes-5302017/.

Cornwall, W. (2007, November 2). Climate can't wait for feds, say Clinton, Gore, Nickels. *Seattle Times*. Retrieved from https://www.seattletimes.com/seattle-news/climate-cant-wait-for-feds-say-clinton-gore-nickels/.

Coven, J. (2020). *Proposed Budget: 2020*. Seattle: Seattle Office of Sustainability and Environment. Retrieved from http://www.seattle.gov/Documents/Departments/FinanceDepartment/20proposedbudget/OSE.pdf.

Cowell, R., & Owens, S. (2010). Revisiting ... governing space: Planning reform and the politics of sustainability. *Environment and Planning C-Government and Policy, 28*(6), 952–57.

Crowley, W., & Wilma, D. (2010). *Power for the People: A History of Seattle City Light*. Seattle: University of Washington Press.

Curtis, S. (ed.) (2014). *The Power of Cities in International Relations*. New York: Routledge.

Cutter, S., Solecki, L., Bragado, N., Carmin, J., Fragkias, M., Ruth, M., & Wilbanks, T. (2014). Urban systems, infrastructure, and vulnerability. Climate change impacts in the United States. In M. Melillo, T. Richmond, & G. Yohe (eds.), *The Third National Climate Assessment*, pp. 282–96. Washington, DC: US Global Change Research Program.

Dalby, S. (2014). Environmental geopolitics in the twenty-first century. *Alternatives: Global, Local, Political, 39*(1), 3–16. doi:10.1177/0304375414558355.

Daly, H. E. (1973). *Toward a Steady-State Economy*. San Francisco: W. H. Freeman.

Danaher, J., Hogan, M. J., Noone, C., Kennedy, R., Behan, A., de Paor, A., ... Shankar, K. (2017). Algorithmic governance: Developing a research agenda through the power of collective intelligence. *Big Data & Society, 4*(2). doi:10.1177/2053951717726554.

David Suzuki Foundation. (2020). Impacts of climate change. Retrieved from https://davidsuzuki.org/what-you-can-do/impacts-climate-change/?gclid=EAIaIQobChMIt Mm1pvih6QIVEz2tBh1tMwmYEAAYAiAAEgLwk_D_BwE.

Davidson, K., & Gleeson, B. (2015). Interrogating urban climate leadership: Toward a political ecology of the C40 network. *Global Environmental Politics, 15*(4), 21–38. doi:10.1162/GLEP_a_00321.

Davoudi, S., Crawford, J., Mehmood, A., Davoudi, S., Crawford, J., & Mehmood, A. (2009). *Planning for Climate Change: Strategies for Mitigation and Adaptation for Spatial Planners*. London: Earthscan.

Day, J. W., & Hall, C. (2016). *America's Most Sustainable Cities and Regions: Surviving the 21st Century Megatrends*, 1st ed.. New York: Springer.

Deetjen, T. A., Conger, J. P., Leibowicz, B. D., & Webber, M. E. (2018). Review of climate action plans in 29 major U.S. cities: Comparing current policies to research recommendations. *Sustainable Cities and Society, 41*, 711–27. doi:https://doi.org/10.1016/j.scs.2018.06.023.

Deng, D., Zhao, Y., & Zhou, X. (2017). Smart city planning under the climate change condition. *IOP Conference Series: Earth and Environmental Science, 81*, 012091. doi:10.1088/1755-1315/81/1/012091.

Department of Finance and Administrative Services. (2019). *Green Fleet Action Plan: An Updated Action Plan for the City of Seattle.* Seattle: City of Seattle. Retrieved from https://www.seattle.gov/Documents/Departments/FAS/FleetManagement/2019-Green-Fleet-Action-Plan.pdf.

Dev, J., & Brazile, L. (2019, March 28). In Seattle, school segregation is actually getting worse. Retrieved from https://crosscut.com/2019/03/seattle-school-segregation-actually-getting-worse.

Dierwechter, Y. (2008). *Urban Growth Management and Its Discontents: Promises, Practices and Geopolitics in US City-Regions.* New York: Palgrave.

———. (2010). Metropolitan geographies of US climate action: Cities, suburbs and the local divide in global responsibilities. *Journal of Environmental Policy and Planning, 12*(1), 59–82.

———. (2013). Smart city-regionalism across Seattle: Progressing transit nodes in labor space? *Geoforum, 49*, 139–49.

———. (2014). The spaces that smart growth makes: Sustainability, segregation, and residential change across Greater Seattle. *Urban Geography, 35*(5), 691–714. doi:10.1080/02723638.2014.916905.

———. (2017). *Urban Sustainability through Smart Growth: Intercurrence, Planning, and the Geographies of Regional Development across Greater Seattle.* Cham, Switzerland: Springer.

———. (2018). What's smart growth got to do with smart cities? Searching for the "Smart City Region" in Greater Seattle. *Territorio, 83*, 48–55.

———. (2019). *The Urbanization of Green Internationalism.* New York: Palgrave.

Dierwechter, Y., & Pendras, M. (2020). Keeping blue collars in green cities: From TOD to TOM? *Frontiers in Sustainable Cities, 2*(7), 1–6. doi:10.3389/frsc.2020.00007.

Dierwechter, Y., & Wessells, A. (2013). The uneven localisation of climate action in metropolitan Seattle. *Urban Studies, 50*(7), 1368–85. doi:10.1177/0042098013480969.

Dilworth, R. (ed.) (2009). *The City in American Political Development.* New York: Routledge.

Dodman, D., & Satterthwaite, D. (2008). Institutional capacity, climate change adaptation and the urban poor. *IDS Bulletin, 39*(4), 67–74. doi:10.1111/j.1759-5436.2008.tb00478.x.

Dooling, S. (2009). Ecological gentrification: A research agenda exploring justice in the city. *International Journal of Urban and Regional Research, 33*(3), 621–39. doi:10.1111/j.1468-2427.2009.00860.x.

Doyle, M. W., Stanley, E. H., Havlick, D. G., Kaiser, M. J., Steinbach, G., Graf, W. L., … Riggsbee, J. A. (2008). Aging infrastructure and ecosystem restoration. *Science, 319*(5861), 286–88.

Dreier, P. (2013, December 19). Radicals in City Hall: An American tradition. *Dissent.* Retrieved from https://www.dissentmagazine.org/online_articles/radicals-in-city-hall-an-american-tradition.

Dreier, P., Mollenkopf, J. H., & Swanstrom, T. (2014). *Place Matters: Metropolitics for the Twenty-First Century*, 3rd ed. Lawrence, KS: University Press of Kansas.

Fainstein, S. (1999). Can we make the cities that we want? In S. Body-Gendrot & R. Beuaregard (eds.), *The Urban Moment.* Thousand Oaks, CA: Sage.

Fesler, S. (2019, January 24). Looking back on three years of transit investment in Seattle. Retrieved from https://www.theurbanist.org/2019/01/24/looking-back-on-three-years-of-transit-investment-in-seattle/.

REFERENCES

Fisher, J. (2020). Reference Archivist, Seattle Municipal Archives. Personal communication, February 20–26 and March 4.

Ford, K. (2010). *The Trouble with City Planning: What New Orleans Can Teach Us*. New Haven, CT: Yale University Press.

Foster, J., Clark, B., & York, R. (2010). *The Ecological Rift: Capitalism's War on the Earth*. New York: Monthly Review Press.

Fowler, C. (2016). Segregation as a multiscalar phenomenon and its implications for neighborhood-scale research: The case of South Seattle 1990–2010. *Urban Geography*, *37*(1), 1–25.

Futurewise. (2017). *Seattle Climate Challenges Atlas*. Retrieved from http://www.futurewise.org/assets/card_images/Futurewise-Climate-Challenges-Atlas-2017.pdf.

Gately, C., Hutyra, L. R., & Wing, I. S. (2019). DARTE Annual On-road CO2 Emissions on a 1-km Grid, Conterminous USA V2, 1980–2017: ORNL Distributed Active Archive Center.

Geels, F. W. (2011). The multi-level perspective on sustainability transitions: Responses to seven criticisms. *Environmental Innovation and Societal Transitions*, *1*(1), 24–40. doi:10.1016/j.eist.2011.02.002.

Georgiadis, N. (2020). Center for Urban Waters & Puget Sound Institute, Tacoma. Personal communication, March 3.

Gerring, J. (2003). APD from a methodological point of view. *Studies in American Political Development*, *17*(1), 82–102.

Gibson, T. (2004). *Securing the Spectacular City: The Politics of Revitalization and Homelessness in Downtown Seattle*. Lanham, MD: Lexington Books.

Glasmeier, A., & Christopherson, S. (2015). Thinking about smart cities. *Cambridge Journal of Regions, Economy and Society*, *8*(1), 3–12. doi:10.1093/cjres/rsu034.

Global Covenant of Mayors. (2018). *Common Reporting Framework, Version 6.1*. Retrieved from https://www.globalcovenantofmayors.org/wp-content/uploads/2019/04/FINAL_Data-TWG_Reporting-Framework_website_FINAL-13-Sept-2018_for-translation.pdf.

Godschalk, D. (2004). Land use planning challenges: Coping with conflicts in visions of sustainable development and livable communities. *Journal of the American Planning Association*, *70*(1), 5–13. doi:10.1080/01944360408976334

Godschalk, D., Rodríguez, D., Berke, P., & Kaiser, E. (2006). *Urban Land Use Planning*, 5th ed. Urbana: University of Illinois Press.

GotGreen! & Puget Sound Sage. (2016). *Our People, Our Planet, Our Power*. Retrieved from https://gotgreenseattle.org/wp-content/uploads/2016/03/OurPeopleOurPlanetOurPower_GotGreen_Sage_Final1.pdf.

Grafakos, S., Pacteau, C., Delgado, M., Landeur, M., Lucon, O., & Driscoll, P. (2018). Integrating mitiation and adaptation: Opportuniites and challenges. In C. Rosenzweig, W. Solecki, P. Romero Lankao, S. Mehrotra, S. Dhakal, & S. Ibrahim (eds.), *Climage Change and Cities*. Cambridge: Cambridge University Press.

Gray, M., Golob, E., & Markusen, A. (1996). Big firms, long arms, wide shoulders: The "hub-and-spoke" industrial district in the Seattle region. *Regional Studies*, *30*(7), 651–66. doi:10.1080/00343409612331349948.

Gregory, J. (2015). Seattle's left coast formula. *Dissent*, *62*(1), 64–70.

Haarstad, H. (2017). Constructing the sustainable city: Examining the role of sustainability in the "smart city" discourse. *Journal of Environmental Policy & Planning*, *19*(4), 423–37. doi:10.1080/1523908X.2016.1245610.

Haskins, S., Gale, D., & Kelly, L. (2002). Creating and optimizing new forms of public-private partnerships in Seattle. *Water Supply*, *2*(4), 211218. doi:10.2166/ws.2002.0140.

Haupt, W., & Coppola, A. (2019). Climate governance in transnational municipal networks: Advancing a potential agenda for analysis and typology. *International Journal of Urban Sustainable Development*, *11*(2), 123–40. doi:10.1080/19463138.2019.1583235.

Hays, S. P. (1959). *Conservation and the Gospel of Efficiency: The Progressive Conservation Movement, 1890–1920*: Cambridge: Harvard University Press.

Healey, P. (2006). *Collaborative Planning: Shaping Places in Fragmented Societies*. Basingstoke, Hampshire: Palgrave Macmillan.

Heikkinen, M., Karimo, A., Klein, J., Juhola, S., & Ylä-Anttila, T. (2020). Transnational municipal networks and climate change adaptation: A study of 377 cities. *Journal of Cleaner Production*, *257*, 120474. doi:10.1016/j.jclepro.2020.120474.

Herrschel, T., & Dierwechter, Y. (2018). *Smart Transitions in City Regionalism: Territory, Politics and the Quest for Competitiveness and Sustainability*. New York: Routledge.

Herrschel, T., & Newman, P. (2017). *Cities as International Actors: Urban and Regional Governance beyond the Nation State*. London: Palgrave.

Hess, T. (2019). *The Role of Cities in Global Climate Governance—The Case of Berlin, Hamburg, and Munich* (master's thesis), University of Lund, Lund. Retrieved from https://lup.lub.lu.se/student-papers/search/publication/8997737.

Hillman City Colloboratory. (2019). Hillman City Collaboratory: An incubator for social change. Retrieved from https://hillmancitycollaboratory.org/.

Himes, D. (2019). Inequality and metropolitan areas. *Monthly Labor Review*, August, https://link.gale.com/apps/doc/A601908857/ITOF?u=wash_main&sid=ITOF&xid=b237a930.

Hinshaw, M. (2016). Public lecture. University of Washington, Tacoma, February 25.

Hollands, R. G. (2008). Will the real smart city please stand up? *City*, *12*(3), 303–20. doi:10.1080/13604810802479126.

———. (2015). Critical interventions into the corporate smart city. *Cambridge Journal of Regions, Economy and Society*, *8*(1), 61–77. doi:10.1093/cjres/rsu011.

Holmes, J. (2020) City of Seattle. Interview (conducted by Carlson Doug and Chris Moradi). April 16.

Honachefsky, W. B. (2000). *Ecologically Based Municipal Land Use Planning*. Boca Raton, FL: Lewis.

Hughes, S. (2016). The politics of urban climate change policy: Toward a research agenda. *Urban Affairs Review*, *53*(2), 362–80. doi:10.1177/1078087416649756.

Inha, L. M., & Hukka, J. J. (2019). Policies enabling resilience in Seattle's water services. *European Journal of Creative Practices in Cities and Landscapes*, *2*(1), 28. doi:10.6092/issn.2612-0496/8688.

Inouye, A. (2020). Public lecture. Director of Strategic Partnerships, Office of Mayor Jacob Frey, City of Minneapolis, UW Tacoma. March 4.

Irwin, E., et al. (2019). The role of regulations and norms in land use change. In S. Pickett, M. Cadenasso, J. Grove, E. Irwin, E. Rosi, & C. Swan (eds.), *Science for the Sustainable City: Empirical Insights from the Baltimore School of Urban Ecology*. New Haven, CT: Yale University Press.

Isaksen, T. B., Yost, M. G., Hom, E. K., Ren, Y., Lyons, H., & Fenske, R. A. (2015). Increased hospital admissions associated with extreme-heat exposure in King County, Washington, 1990–2010. *Reviews on Environmental Health*, *30*(1), 51. doi:10.1515/reveh-2014-0050.

Ivanova, M. (2010). UNEP in global environmental governance: Design, leadership, location. *Global Environmental Politics, 10*(1), 30–59. doi:10.1162/glep.2010.10.1.30.

Jackson, J., Yost, M., Karr, C., Fitzpatrick, C., Lamb, B., Chung, S., ... Fenske, R. (2010). Public health impacts of climate change in Washington state: Projected mortality risks due to heat events and air pollution. *Climatic Change, 102*(1–2), 159–86. doi:10.1007/s10584-010-9852-3.

Jacobs, J. (1969). *The Economy of Cities*. New York: Random House.

Janos, N. (2018). Urbanising territory: The contradictions of eco-cityism at the industrial margins, Duwamish River, Seattle. *Urban Studies, 57*(11), 2282–99. doi:10.1177/0042098018797284.

Janos, K., & McKendry, C. (2014). Globalization, governance, and renaturing the industrial city: Chicago, IL, and Seattle, WA. In S. Curtis (ed.), *The Power of Cities in International Relations*. London: Routledge.

Johnson, C. (2018). *The Power of Cities in Global Climate Politics*. New York: Palgrave.

Johnson, R. L., & Staeheli, P. (2006). City of Seattle: Stormwater low impact development practices. *World Environmental and Water Resource Congress 2006*, pp. 1–10.

Jonas, A. (2020). The new urban managerialism in geopolitical context. *Dialogues in Human Geography, 10*(3), 330–35. doi:10.1177/2043820620921031.

Jones, B. (2020). Washington Field Organizer, Climate Solutions, Seattle. Personal communication, July 2.

Kantor, P. (1988). *The dependent city: The changing political economy of urban America*. Glenview, IL: Scott Foresman.

Karvonen, A. (2011). *Politics of Urban Runoff: Nature, Technology, and the Sustainable City*. Cambridge, MA: MIT Press.

Kessler, R. (2011). Stormwater strategies: Cities prepare aging infrastructure for climate change. *Environmental Health Perspectives, 119*(12), a514–19. doi:10.1289/ehp.119-a514.

King County. (1958). *Comprehensive Sewage Plan. Natural Resources and Parks*. Seattle: Wastewater Treatment Dept, .

———. (2011). *SEPA GHG Emissions Worksheet Version 1.7 12/26/07*. Seattle: Department of Development and Environmental Service. Retrieved from https://www.kingcounty.gov/~/media/depts/permitting-environmental-review/dper/documents/PublicNotices/ELEC19-0029_SEPA-checklist-gas-emissions-worksheets.ashx?la=en.

King County–Cities Climate Collaboration. (2018). *King County Parks Solar Energy Fact Sheet*. Seattle: Department of Natural Resrouces and Parks. Retrieved from https://your.kingcounty.gov/dnrp/climate/documents/2018-K4Cfact-KC-parks-solar.pdf.

———. (2020). *Recap of Recent Climate and Energy State Action and K4C Interests for 2020*. Seattle: Department of Natural Resources and Park. Retrieved from https://your.kingcounty.gov/dnrp/climate/documents/2020-K4C-Legislative-Interests.pdf.

Kirkendall, R. S. (1994). The Boeing Company and the military-metropolitan-industrial complex, 1945–1953. *Pacific Northwest Quarterly, 85*(4), 137–49. doi:10.2307/40491582.

Klauber, A., Benn, A., Hardenbol, C., Schiller, C., Toussie, I., Valk, M., & Waller, J. (2017). *Innovative Funding for Sustainable Aviation Fuel at U.S. Airports: Explored at Seattle-Tacoma International*. Rocky Mountain Institute. Retrieved from https://www.rmi.org/insights/reports/innovativefunding-sea-tac-2017/.

Klingle, M. (2007). *Emerald City: An Environmental History of Seattle*. New Haven, CT: Yale University Press.

Knieling, J. (2016). *Climate Adaptation Governance in Cities and Regions: Theoretical Fundamentals and Practical Evidence*. Chichester, UK: Wiley Blackwell.

Kolbert, E. (2014). *The Sixth Extinction: An Unnatural History*. New York: Henry Holt.
Kristin, L. (2015). *The Global City 2.0: From Strategic Site to Global Actor*. London: Taylor and Francis.
Krueger, R., & Agyeman, J. (2005). Sustainability schizophrenia or "actually existing sustainabilities?" Toward a broader understanding of the politics and promise of local sustainability in the US. *Geoforum, 36*(4), 410–17. doi:10.1016/j.geoforum.2004.07.005.
Kylili, A., & Fokaides, P. A. (2015). European smart cities: The role of zero energy buildings. *Sustainable Cities and Society, 15*, 86–95. doi:10.1016/j.scs.2014.12.003.
Laurian, L., Walker, M., & Crawford, J. (2017). Implementing environmental sustainability in local government: The impacts of framing, agency culture, and structure in US cities and counties. *International Journal of Public Administration, 40*(3), 270–83. doi:10.1080/01900692.2015.1107738.
Lawrence, G. (1996). Sustainability in Seattle. *City, 1*(3–4), 111–21. doi:10.1080/13604819608713437.
Lee, T., & Painter, M. (2015). Comprehensive local climate policy: The role of urban governance. *Urban Climate, 14*, 566–77. doi:10.1016/j.uclim.2015.09.003.
Lee, T., & van de Meene, S. (2012). Who teaches and who learns? Policy learning through the C40 cities climate network. *Policy Sciences, 45*(3), 199–220. doi:10.1007/s11077-012-9159-5.
Lieberherr-Gardiol, F. (2008). Urban sustainability and governance: Issues for the twenty-first century. *International Social Science Journal, 59*(193–194), 331–42. doi:10.1111/j.1468-2451.2009.01670.x.
Liptan, T. (2017). *Sustainable Stormwater Management: A Landscape-Driven Approach to Planning and Design*. Portland: Timber Press.
Little, A. (2005). An interview with Seattle Mayor Greg Nickels on his pro-Kyoto cities initiative. Retrieved from https://grist.org/article/little-nickels/.
Long, K. (2019, August 27). Seattle home prices again lower than a year ago. *Seattle Times*. Retrieved from https://www.seattletimes.com/business/real-estate/seattle-home-prices-lower-than-a-year-ago-for-second-month-in-a-row/.
Lorente-Plazas, R., Mitchell, T. P., Mauger, G., & Salathé, E. P. (2018). Local enhancement of extreme precipitation during atmospheric rivers as simulated in a regional climate model. *Journal of Hydrometeorology, 19*(9), 1429–46. doi:10.1175/JHM-D-17-0246.1.
Lowe, K. (2014). Bypassing equity? Transit investment and regional transportation planning. *Journal of Planning Education and Research, 34*(1), 30–44. doi:10.1177/0739456x13519474.
Luis, M. (2012). *Century 21 City: Seattle's Fifty Year Journey from World's Fair to World Stage*. Medina, WA: Fairweather.
Lyons, J. (2004). *Selling Seattle: Representing Contemporary Urban America*. London: Wallflower.
MacDonald, N. (1987). *Distant Neighbors: A Comparative History of Seattle & Vancouver*. Seattle: University of Washington Press.
Mason, S. G., & Fragkias, M. (2018). Metropolitan planning organizations and climate change action. *Urban Climate, 25*, 37–50. doi:10.1016/j.uclim.2018.04.004.
Mathias, A. (2019). City of Renton. Personal communication, August 1.
Mathiesen, K. (2017). Seattle pledges support for climate fund barred by Trump. *Climate Change News*. Retrieved from https://d1wqtxts1xzle7.cloudfront.net/53643998/Seattle_pledges_support_for_climate_fund_barred_by_Trump___Climate_Home_-_climate_change_news.pdf?1498.
Mauger, G., Hegewisch, K., Miller, I., Won, L., & Lynch, C. (2018a). Global climate projections, sea level rise, and extreme precipitation. Final report to the Port Gamble S'Klallam Tribe, Climate Impacts Group, University of Washington, Seattle.

Mauger, G., Won, J., Hegewisch, K., Lynch, C., Lorente Plazas, R., & Salathé. E. (2018b). New projections of changing heavy precipitation in King County. Report prepared for the King County Department of Natural Resources, Climate Impacts Group, University of Washington, Seattle.

Mayer, H. (2013). Entrepreneurship in a hub-and-spoke industrial district: Firm survey evidence from Seattle's technology industry. *Regional Studies, 47*(10), 1715–33. doi:10.1080/00343404.2013.806792.

Mazower, M. (2012). *Governing the World: The History of an Idea, 1815 to the Present.* New York: Penguin.

McCormick, K. (2017). Make no small plans: The evolution of comprehensive plans. *Land Lines*, April, 18–28.

McGourtney, K. (2020). Director of Transportation Planning, Puget Sound Regional Counicl, Seattle. Personal communication, July 13.

McIntyre, M. (2014). *Chamber Businesses Support Climate Change Declaration.* Retrieved from https://www.seattlechamber.com/home/advocacy/advocacy-news/details/2014/10/27/chamber-businesses-support-climate-change-declaration.

McKean, G. L. (1941). Tacoma, lumber metropolis. *Economic Geography, 17*(3), 311–20.

McLaren, D., & Agyeman, J. (2018). Sharing cities for a smart and sustainable future. In T. Haas & H. Westlund (eds.), *The Post-Urban World*, 1st ed., vol. 1, pp. 322–35). London: Routledge.

Miller, I., Morgan, H., Mauger, G., Newton, T., Weldon, R., Schmidt, D., ... Grossman, E. (2018). Projected Sea Level Rise for Washington State—A 2018 Assessment. A collaboration of Washington Sea Grant, University of Washington Climate Impacts Group, Oregon State University, University of Washington, and US Geological Survey. Prepared for the Washington Coastal Resilience Project, Seattle.

Miro, C. R., & Cox, J. E. (1999). Climate change—Seattle issues update seminar. *Ashrae Journal, 41*(9), 25–26.

Moisio, S., & Jonas, A. (2018). City-regions and city-regionalism. In A. Paasi, J. Harrison, & M. Jones (eds.), *Handbook on the Geographies of Regions and Territories.* Cheltenham, UK: Edward Elgar.

Montgomery, M., & Mighetto, L. (1988). *Hard Drive to the Klondike: Promoting Seattle during the Gold Rush: A Historic Resource Study for the Seattle Unit of the Klondike Gold Rush National Historical Park.* Seattle: National Park Service, Columbia Cascades Support Office.

Morandi, C., Rolando, A., & di Vita, S. (2016). *From Smart City to Smart Region: Digital Services for an Internet of Places.* Cham, Switzerland: Springer.

Morgan, M. (1951). *Skid Road: An Informal Portrait of Seattle.* New York: Viking.

Mossner, S., & Miller, B. (2015). Sustainability in one place? Dilemmas of sustaainbility governance in the Frieburg metropolitan region. *Regions, 300*(Winter), 19–21.

Næss, P. (1989). Sustainable urban development: The challenges of the Brundtland commission—a turning point for urban planning? *Scandinavian Housing and Planning Research, 6*(1), 45–49. doi:10.1080/02815738908730179.

Nelson, D. (2011). Adaptation and resilience: Responding to a changing climate. *WIREs Climate Change, 2*, 113–20. doi:10.1002/wcc.91.

Nesbit, R. (1952). Skid road: An informal portrait of Seattle, Murray Morgan. *Pacific Northwest Quarterly, 43*(3), 235–36.

Notre Dame Global Adaptation Initiative. (2018). *Urban Adaptation Assessment Technical Document.* University of Notre Dame, South Bend, IN. Retrieved from https://gain.nd.edu/assets/293226/uaa_technical_document.pdf.

NYT Editorial Board. (2020, April 11). How to save black and Hispanic lives in a pandemic. *New York Times*. Retrieved from https://www.nytimes.com/2020/04/11/opinion/coronavirus-poor-black-latino.html.

Ochsner, J., & Andersen, D. (2002). Meeting the danger of fire: Design and construction in Seattle. *Pacific Northwest Quarterly*, *93*(3), 115–26.

Office of Sustainability and Environment. (2002–2005). *Urban Sustainability Advisory Panel Records (Vol. Box: 7905-01)*. Seattle: Seattle Municipal Archives.

———. (2016a). *Equity and Environment Agenda*. Seattle: City of Seattle. Retrieved from https://www.seattle.gov/Documents/Departments/OSE/SeattleEquityAgenda.pdf.

———. (2016b). *Seattle Tree Canopy Assessment*. Seattle: City of Seattle. Retrieved from http://www.seattle.gov/Documents/Departments/Trees/Mangement/Canopy/Seattle2016CCAFinalReportFINAL.pdf.

———. (2017). *Carbon Neutral, Climate Ready*. Seattle: City of Seattle. Retrieved from https://www.seattle.gov/Documents/Departments/Environment/ClimateChange/SEAClimatePreparedness_August2017.pdf.

Office of the Mayor. (2016). *Executive Order 2016-06*. Seattle: City of Seattle. Retrieved from http://murray.seattle.gov/wp-content/uploads/2016/07/Executive-Order-2016-06.pdf.

Orren, K., & Skowronek, S. (1996). Institutions and intercurrence: Theory building in the fullness of time. *Nomos XXXVII, Political Order*, *38*, 111–46.

———. (2004). *The Search for American Political Development*. Cambridge: Cambridge University Press.

Oseland, S. E. (2019). Breaking silos: Can cities break down institutional barriers in climate planning? *Journal of Environmental Policy & Planning*, *21*(4), 345–57. doi:10.1080/1523908X.2019.1623657.

Osofsky, H. (2015). Rethinking the geography of local climate action: Multilevel network participation in metropolitan regions. *Utah Law Review*, *15*(1), 173–240.

Ott, J. (2008). First National Conservation Congress opens at the Alaska-Yukon-Pacific Exposition on August 26, 1909. *HistoryLink.org, Essay 8775*.

Pacino, V. (2014, September 10). Preparing Seattle for climate change: Lessons learned from adaptation at the local level. Paper presented at the Fifth Annual PNW Climate Science Conference, Seattle, WA.

Parolek, D. (2016). Missing middle housing: Supplying diverse housing options along a spectrum of affordability. *Journal of Case Study Research: A Publication of the Center for California Real Estate*, *1*(1), 32–35.

Pastor, M., Benner, C., Rosner, R., Matsuoka, M., & Jacobs, J. (2004). Community building, community bridging: Linking neighborhood improvement initiatives and the new regionalism in the San Franciscio Bay Area. Santa Cruz. CA: UC Santa Cruz, Center for Justice, Tolerance, and Community.

Peattie, K., & Hall, G. (1994). The greening of local government: A survey. *Local Government Studies*, *20*(3), 458–85. doi:10.1080/03003939408433739.

Pelling, M. (2011). *Adaptation to Climate Change: From Resilience to Transformation*. London: Routledge.

Perry, D. (1995). *Building the Public City: The Politics, Governance, and Finance of Public Infrastructure*. Newbury, CA: Sage.

Pettibone, L. (2015). *Governing Urban Sustainability: Comparing Cities in the USA and Germany*. Farnham, Surrey: Ashgate.

Pickett, S., Cadenasso, M., Grove, J., Irwin, E., Rosi, E., & Swan, C. (eds.). (2019). *Science for the Sustainable City: Empirical Insights from the Baltimore School of Urban Ecology*. New Haven, CT: Yale University Press.

Piketty, T. (2014). *Capital in the Twenty-First Century*. Cambridge, MA: Harvard University Press.

Pollini, J. (2013). Bruno Latour and the ontological dissolution of nature in the social sciences: A critical review. *Environmental Values, 22*, 25–42.

Port of Seattle. (2016). *Seattle-Tacoma International Airport: Strategy for a Sustainable Sea-Tac*. Seattle: Port of Seattle.

Portney, K. E. (2003). *Taking Sustainability Seriously: Economic Development, the Environment, and Quality of Life in American Cities*. Boston: MIT Press.

Prosperity Partnership. (2012). *Regional Economic Strategy for the Central Puget Sound Region Economy*. Retrieved from https://www.psrc.org/our-work/regional-economic-strategy.

Puget Sound Regional Council. (1998). *Regional Review: Monitoring Change in the Central Puget Sound Region*. Seattle: Puget Sound Regional Council.

———. (2019a). *Region 2050: Climate Change: Background Paper*. Seattle. Retreived from https://www.psrc.org/sites/default/files/vision2050climatepaper.pdf.

———. (2019b). *VISION 2050—DRAFT Climate Change Chapter*. Seattle: Puget Sound Regional Council. Retrieved from https://www.psrc.org/sites/default/files/mpp_climate_policy_matrix_7-03-19.pdf.

Puget Sound Sage. (2012). *Ensuring Transit Investment in Seattle's Rainier Valley: Build Communities Where All Families Thrive*. Seattle: Puget Sound Sage.

Purcell, M. (2008). *Recapturing Democracy*. New York: Routledge.

Ramin, B., & Svoboda, T. (2009). Health of the homeless and climate change. *Journal of Urban Health, 86*(4), 654–64.

Ramsey, K. (2009). *Adapting (to) the "Climate Crisis": Urban Environmental Governance and the Politics of Mobility in Seattle* (doctoral dissertation), University of Washington, Seattle.

Raven, J., Stone, B., Mills, G., Towers, J., Katzschner, L., Leone, M., Gaborit, P., Georgescu, M., & Hariri, M. (2018). Urban planning and design. In C. Rosenzweig, W. Solecki, P. Romero-Lankao, S. Mehrotra, S. Dhakal, and S. Ali Ibrahim (eds.), *Climate Change and Cities: Second Assessment Report of the Urban Climate Change Research Network*, pp. 139–72. New York: Cambridge University Press.

Reams, M. A., Clinton, K. W., & Lam, N. S. N. (2012). Achievement of climate planning objectives among U.S. member cities of the International Council for Local Environmental Initiatives (ICLEI). *Low Carbon Economy, 3*(4), 137–43. doi:10.4236/lce.2012.34018.

Reum, J. C. P., Essington, T. E., Greene, C. M., Rice, C. A., & Fresh, K. L. (2011). Multiscale influence of climate on estuarine populations of forage fish: The role of coastal upwelling, freshwater flow and temperature. *Marine Ecology Progress Series, 425*, 203–15.

Rice, J. (2010). Climate, carbon, and territory: Greenhouse gas mitigation in Seattle, Washington. *Annals of the Association of American Geographers, 100*(4), 929–37. doi:10.1080/00045608.2010.502434.

———. (2018). Climate science and the city: Consensus, calculation and security in Seattle, Washington. In K. Ward, A. Jonas, B. Miller, & D. Wilson (eds.), *The Routledge Handbook on Spaces of Urban Politics*. London: Routledge

Rice, J., Cohen, D., Long, J., & Jurjevich, J. (2019). Contradictions of the climate-friendly city: New perspectives on eco-gentrification and housing justice. *International Journal of Urban and Regional Research, 44*(10), 145–65. doi:10.1111/1468-2427.12740.

Ritchie, H., & Roser, M. (2017). CO2 and greenhouse gas emmisions. Retrieved from https://ourworldindata.org/co2-and-other-greenhouse-gas-emissions.

Robinson, L., Newell, J. P., & Marzluff, J. M. (2005). Twenty-five years of sprawl in the Seattle region: Growth management responses and implications for conservation. *Landscape and Urban Planning, 71*(1), 51–72.

Rosdil, D. L. (2017). The survival of progressive urban politics amid economic adversity. *Journal of Urban Affairs, 39*(2), 205–24. doi:10.1111/juaf.12311.

Rosenzweig, C., Solecki, W., Romero Lankao, P., Mehrotra, S., Dhakal, S., & Ibrahim, S. A. (2018). *Climate Change and Cities: Second Assessment Report of the Urban Climate Change Research Network*. New York: Cambridge University Press.

Roy, J. (2017). Smart cities in Canada: An examination of progress and impediments in Halifax, Canada. *International Journal of Services Technology and Management, 23*(5/6), 361–80.

Rush, J. (2016). Public lecture. Forterra. UW Tacoma. February 25.

Rutland, T., & Aylett, A. (2008). The work of policy: Actor networks, governmentality, and local action on climate change in Portland, Oregon. *Environment and Planning D: Society and Space, 26*(4), 627–46. doi:10.1068/d6907.

Salathé, E., Leung, L., Qian, Y., & Zhang, Y. (2010). Regional climate model projections for the State of Washington. *Climatic Change, 102*(1–2), 51–75. doi:10.1007/s10584-010-9849-y.

Sale, R. (1976). *Seattle, Past to Present*. Seattle: University of Washington Press.

Sanders, J. (2010). *Seattle and the Roots of Urban Sustainability: Inventing Ecotopia*. Pittsburgh: University of Pittsburgh.

Schmidt, E. (2019, February 27). I used to run Google. Silicon Valley could lose to China, Opinion. *New York Times*. Retrieved from https://www.nytimes.com/2020/02/27/opinion/eric-schmidt-ai-china.html?referringSource=articleShare.

Schragger, R. C. (2013). Is a progressive city possible? Reviving urban liberalism for the twenty-first century. *Harvard Law & Policy Review, 7*(2), 231–52.

Scott, A. (2001). Globalization and the rise of city regions. *European Planning Studies, 9*(7), 813–26.

———. (2011). A world in emergence: Notes toward a resynthesis of urban-economic geography for the 21st century. *Urban Geography, 36*(2), 845–70.

Scott, K. (2016). *The Digital City and Mediated Urban Ecologies*. Cham, Switzerland: Palgrave Macmillan.

Seattle Board of Public Works. (1918). *Specifications for Complete Hydro-Electric Plant, Being an Extension to the Municipal Electric Light and Power System of Seattle*. Seattle: Lowman & Hanford.

Seattle Office for Civil Rights. (2008). *Inclusive Outreach and Public Engagement Guide*. Seattle: City of Seattle.

Seattle Public Utilities. (2007). *Water Services Plan*. Seattle: Seattle Public Utilities.

———. (2019). *SPU Equity Planning Toolkit: Stakeholder Analysis*. Retrieved from https://www.seattle.gov/Documents/Departments/SPU//EquityPlanningTools.pdf.

Sellers, J. M. (2002). *Governing from below: Urban Regions and the Global Economy*. Cambridge: Cambridge University Press.

Shaw, R., Pulhin, J., Pereira, J., & Pereira, J. J. (2010). *Climate Change Adaptation and Disaster Risk Reduction: Issues and Challenges*. Bingley, UK: Emerald.

Shelton, T., Zook, M., & Wiig, A. (2015). The "actually existing smart city." *Cambridge Journal of Regions, Economy and Society, 8*(1), 13–25. doi:10.1093/cjres/rsu026.

Söderström, O., Paasche, T., & Klauser, F. (2014). Smart cities as corporate storytelling. *City, 18*(3), 307–20. doi:10.1080/13604813.2014.906716.

Solecki, W., Leichenko, R., & O'Brien, K. (2011). Climate change adaptation strategies and disaster risk reduction in cities: Connections, contentions, and synergies. *Current Opinion in Environmental Sustainability, 3*(3), 135–41. doi:10.1016/j.cosust.2011.03.001.

Soper, T. (2017). Why Seattle is posed to be a leader in smart city technology. Retrieved from https://www.geekwire.com/2017/seattle-poised-leader-establishing-smart-city-regulations-technology/.

Sound Transit. (2018). *RESOLUTION NO. R2018-10: Adopting an Equitable Transit Oriented Development Policy*. Seattle: Sound Transit. Retrieved from https://www.soundtransit.org/st_sharepoint/download/sites/PRDA/FinalRecords/2018/Resolution%20R2018-10.pdf.

Stanton, L. (2017). Public lecture. Port of Seattle. UW Tacoma. March 23.

Steven, J. (2011). May 4, 1969: Hit the highway, freeway. Retrieved from https://radsearem.wordpress.com/2011/05/04/may-4-1969-hit-the-highway-freeway/.

Stewardship Partners. (2020). 5th Annual Puget Sound Green Infrastructure Summit. Retrieved from https://www.12000raingardens.org/%20summit/.

Sustainable Capitol Hill. (2020). Sustainable Capitol Hill Tool Project: Build your own community. Retrieved from https://sustainablecapitolhill.org/tool-library/.

Tackett, T. (2008). Street alternatives: Seattle Public Utilities' natural drainage system program. In M. Clar (ed.), *Low Impact Development: New and Continuing Applications*, pp. 316–21. Reston, VA: American Society of Civil Engineers.

Tate, C. (2000). Voters elect People's Ticket candidate Beriah Brown as mayor of the City of Seattle on July 8, 1878. *HistoryLink.org, Essay 2778*.

Taylor, P., O'Brien, G., & O'Keefe, P. (2020). *Cities Demanding the Earth: A New Understanding of the Climate Emergency*. Bristol: Bristol University Press.

Taylor, Q. (1994). *The Forging of a Black Community: Seattle's Central District, from 1870 through the Civil Rights Era*. Seattle: University of Washington Press.

Thrush, C.-P. (2007). *Native Seattle: Histories of the Crossing over Place*. Seattle: University of Washington Press.

Tigue, K. (2019, August 15). What Would a City-Level Green New Deal Look Like? Seattle's About to Find Out. Retrieved from https://insideclimatenews.org/news/13082019/seattle-city-green-new-deal-heating-oil-tax-free-public-transit-congestion-pricing-resolution.

Tretter, E., & Mueller, E. (2018). Transforming Rainey Street: The decoupling of equity from environment in Austin's smart growth agenda. In K. Ward et al. (eds.), *The Routledge Handbook on Spaces of Urban Politics*. New York: Routledge.

Trisos, C. H., Merow, C., & Pigot, A. L. (2020). The projected timing of abrupt ecological disruption from climate change. *Nature, 580*(7804), 496. doi:10.1038/s41586-020-2189-9.

UCCRN. (2018). Who we are. Retrieved from http://uccrn.org/who-we-are/overview/.

Uloa, A. (2017). The geopolitics of carbonized nature and the zero carbon citizen. *South Atlantic Quarterly, 116*(1), 111–20. doi:10.1215/00382876-3749359.

UN Framework Convention on Climate Change. (2018). 100+ cities produce more than 70% of electricity from renewables—CDP (press release). Retrieved from https://unfccc.int/news/100-cities-produce-more-than-70-of-electricity-from-renewables-cdp.

UN-Habitat. (2009). *Cities and Climate Change Initiative: Launch and Conference Report*. Oslo: UN-Habitat and Norwegian Ministry of Foreign Affairs.

US Department of Housing and Urban Development. (2018). *The 2018 Annual Homeless Assessment Report (AHAR) to Congress.* Washington, DC: US Department of Housing and Urban Development.

UW Climate Impacts Group, UW Department of Environmental and Occupational Health Sciences, Front and Centered, & Urban@UW. (2018). *An Unfair Share: Exploring the Disproportionate Risks from Climate Change Facing Washington State Communities. A Report Prepared for Seattle Foundation. University of Washington, Seattle.* Retrieved from https://cig.uw.edu/our-work/applied-research/an-unfair-share-report/.

Vincent, C. (2019). City of Renton. Personal communication, August 1.

Wang, R. (2012). Leaders, followers, and laggards: Adoption of the US Conference of Mayors Climate Protection Agreement in California. *Environment and Planning C: Government and Policy, 30*(6), 1116–28. doi:10.1068/c1122.

White House. (2014). FACT SHEET: 16 U.S. communities recognized as climate action champions for leadership on climate change (press release). Retrieved from https://obamawhitehouse.archives.gov/the-press-office/2014/12/03/fact-sheet-16-us-communities-recognized-climate-action-champions-leaders.

Whittemore, A. H. (2015). Practitioners theorize, too: Reaffirming planning theory in a survey of practitioners' theories. *Journal of Planning Education and Research, 35*(1), 76–85. doi:10.1177/0739456X14563144.

Yohe, G., Lasco, R., Ahmad, A., Arnell, N., Cohen, S., Hope, C., ... Perez, R. (2007). Perspectives on climate change and development. In M. Parry, O. Canziani, J. Palutikof, P. van der Linden, & C. Hanson (eds.), *Climate Change 2007: Impacts, Adaptation and Vulnerability. Contribution of Working Group II to the Fourth Assessment Report of the Intergovernmental Panel on Climate Change*, pp. 811–41. Cambridge: Cambridge University Press.

Zahran, S., Brody, S. D., Vedlitz, A., Grover, H., & Miller, C. (2008). Vulnerability and capacity: Explaining local commitment to climate-change policy. *Environment and Planning C-Government and Policy, 26*(3), 544–62. doi:10.1068/c2g.

INDEX

350 Seattle 29, 31

"actually-existing sustainabilities"
  concept of 50
advancing carbon networks 1, 3, 4, 5, 6,
  30, 35, 42, 54, 55–59, 79, 85–86, 91
Airbnb 52
Alaska Way Viaduct 37, 82
Alaska-Yukon World's Fair 37, 48, 94
Amazon 5, 11, 17, 18, 25, 28, 33, 95
ARC3.2, 4, 5, 7, 8, 12, 13, 16, 36, 66, 78,
  79, 88, 91

Barber, Benjamin 30, 50
Bloomberg, Michael 50, 55
blue collars in green cities
  Dierwechter and Pendras on 14, 81, 87
Boeing 10, 11, 17, 18, 21, 26, 96
Brundtland Report 2
Bulkeley, Harriet 94, 96
Burnham Plan 40
Bush (George W) administration 2,
  27, 41, 56

C40 Cities 3, 4, 12, 30, 35, 51, 54
Capitol Hill Eco-District Project 93
Capitol Hill Occupation Protest.
  See CHOP
Carbon Neutral Cities Alliance 12, 35,
  57, 86, 91
coordinating disaster risk reduction with
  climate change adaptation 1, 7, 79,
  81, 83, 91
CHOP 33, 93
climate action plan (2006) 42, 51
climate action plan (2013) 48, 49, 51
climate change 1

and activism 37, 48, 94
and cities 8, 13, 36, 60
and city-regionalism 3, 55, 58–60, 85
and disaster risk reduction 1, 5, 6, 78,
  81–83, 89
global scale 35, 38, 39, 56, 57, 93
impacts of 2, 10, 49, 63, 66, 71, 74, 87
impacts on planning policy 57
and international organizations
  3, 54, 57
origins and causes of 8
and Seattle 62, 66, 68, 69, 82
as a 'pentagon' of issues 5
Climate Impacts Group (UW) 74
climate justice organizations 70, 71
Climate Solutions 29–30
CO_2 equivalent. *See* CO2e
CO2e 6, 44, 45, 55, 56, 60, 62, 68, 71,
  79, 81, 91
cogenerating risk information 1
coordinating disaster risk reduction with
  climate change adaption 1, 7,
  79, 83, 91
COVID-19, 54, 67, 69, 81
  uneven social impacts of 69
coworking facilities 96

defense of privilege thesis 26
Department of Emergency
  Management 82, 83
Department of Finance and
  Administrative Services 80, 93
disadvantaged populations 1, 5, 6, 7, 11,
  66, 79, 86–89, 91
Durkan, Jenny 51, 54
Duwamish River 17, 20
Duwamish River Basin 10, 87

East African Community Services 90
ecological gentrification 10, 46, 81
economic segregation 25
ecosystem services 73
Elite Emerald 1, 10, 46
Emerald City 1, 10, 15, 46, 87, 98
Environmental Justice and Service Equity Division 70
Environmental Professionals of Color 48
Equitable Justice Delivery System 23, 24
equity 2, 4, 6, 7, 10, 22, 24, 26, 32, 33, 37, 40, 41, 48, 49, 53, 54, 70, 73, 74, 76, 78, 84, 86, 87, 89

food sovereignty 62
Freiburg, Germany 28, 58, 59

gentrified sustainability 71
George Floyd
  impact on Seattle 33, 93
Global Covenant of Mayors 3, 12, 35, 64, 57
Got Green! 29, 70, 71, 93, 94
grassroots counterculture 37
green city-regionalism 3, 55, 58–60
Growth Management Act 21, 30, 39, 86, 95

Hidalgo, Anne 50
High Point 76
historic institutionalism 16
homelessness 26, 81, 84, 86
HOPE VI, 76
hydropower 26, 81, 84, 86

ICLEI 2, 4, 55, 57
Industrial Revolution 65
Inslee, Jay 96
integrated planning 80
integrating mitigation with adaptation 1, 78–81, 83, 91
intercurrence
  theoretical concept of 31, 42, 74, 75
International Council for Local Environmental Initiatives. *See* ICLEI
intersectionality 29, 30

kayaktivists 37, 95
K4C 35, 55, 58–60
King County's Fund to Reduce Energy Demand (FRED) 60
King County–Cities Climate Collaboration. *See* K4C
Klondike Gold Rush 17
Kyoto 2, 27, 41–43, 55

LA21 2
Lake Washington 20, 36
  clean up campaign for 20
Local Agenda 21. *See* LA21
local Green New Deal 29, 41, 60, 89, 93
low-impact development 46, 76, 77

market liberalism 4, 19, 50
Mayors Climate Protection Agreement 12, 27, 35, 55–57, 85
McGinn, Mike 70
Metropolitan Municipality of Seattle 20, 75
Microsoft 5, 11, 17, 18, 25, 95
multiple orders
  theoretical concept of 15, 24, 31, 74, 79

National Oceanic and Atmospheric Administration 68
Network for Business Innovation and Sustainability 62
New Urbanism 39
Nickels, Greg 27, 51, 54–56, 85, 93
Northwest Seaport Alliance 59

Obama administration 97
Obama, Barack 42
ocean acidification 32, 58, 68
Office of Planning and Development 49, 52, 56, 70, 74, 84, 93
Office of Sustainability and Envrionment. *See* OSE
OSE 49, 52, 53

Paris Agreement 91, 92
populations of color
  in Seattle 6, 71

INDEX

Port of Seattle 3, 12, 28, 31, 59
   and bio-fuels 29, 96
PSRC 29, 31, 56, 59, 95
public sewerage services
   fragmentation of 20
Puget Sound Clean Air Agency 29, 59
Puget Sound Regional Council. *See* PSRC
Puget Sound Sage 46, 70, 71, 93

Raging Grannies 48, 69
Rio Earth Summit 2

SDGs 2
SEA Streets 76, 97
sea-level rise 32, 49, 58, 67, 71
Sea–Tac International Airport 28, 96
Seattle
   African-American population in 26, 33
   annexation history 20
   as a "boom-and-bust" city 18, 24, 33
   and capacity to spin-off new firms 18
   climate change impacts on 32, 49, 71
   climate change projections for 12, 49, 66, 67, 74
   conflicting pressures in 24, 33
   "dual personality" of 19
   early development of water systems in 20
   economic restructuring in 1980s 25, 40
   grassroots counterculture in 37, 90
   high-tech urban core 28, 33, 61, 81
   and homelessness 26, 81, 84, 86
   housing composition in 25, 26
   main climate challenges in 25, 40
   origins of 11, 16, 17
   political culture in 19, 56, 87
   recent growth in 40
   relationship with San Francisco 10, 12, 16, 25, 56, 57, 68, 86
   smart city developments in 11, 23, 24, 31, 48, 84, 89
   as socio-natural system 16, 69, 76
   support for Paris Agreement 91
   type of local government 12, 27
   in World War I 17
Seattle 2035 32, 40, 41

Seattle City Light 22, 31, 48, 49, 51, 52, 54, 60, 80, 89, 92, 93
Seattle Metropolitan Chamber of Commerce 17, 27, 30, 31, 62
Seattle metropolitan region 6, 10, 13, 37
Seattle Public Utilities. *See* SPU
Seattle Transportation Benefit District 92
Seattle's Renters Commission 84
Seattle–Tacoma–Bellevue metropolitan statistical area 8, 21
sharing city
   concept of 52, 62, 93, 96
Shell, Paul 52
smart cities 11, 23, 24, 31, 48, 84, 85, 89
Smart Growth 30, 39, 40, 45, 85, 89
smart segregation 10, 46, 81
social justice 4, 5, 21, 24, 30, 40, 41, 48, 49, 69, 70, 84, 93, 96. *See also* 'equity'
   impacts on planning policy 83
   Seattle groups advocating for 54, 71, 81, 95
Sound Transit 21, 29, 31, 46, 59, 86, 95
SPU 70, 73–78, 86, 89
stormwater policies 33, 71, 75, 83, 97
Sustainable Capitol Hill 52
sustainable development
   concept of 1, 2, 40
Sustainable Development Goals. *See* SDGs
Sustainable Seattle
   planning theme 31, 39, 40

Tacoma 16, 17, 26, 55, 59
TOD 21, 46, 63, 81, 94
Tolt River 19, 73
   water supply to Seattle 73
Towards a Sustainable Seattle (1994 plan) 39–41
transit-oriented development. *See* TOD
tranational climate action networks: 4, 30, 54, 57, 58, 85, 91

UCCRN 1, 3, 4, 5, 12, 16, 66, 67, 91. *See* Urban Climate Change Research Network
UNFCC 83

United Nations Environment Programme 2, 4
United Nations Human Settlements Programme 2
Urban Climate Change Research Network 1, 3, 4, 5. *See* UCCRN
urban infrastructure 12, 23
  decay of 73
urban sustainability 8, 12, 24, 27, 33, 35, 37–39, 45, 51, 56, 84
urban transformation
  as five pathways of action 4, 5, 16, 78, 79
urban village model 40, 43, 45, 62, 85, 93

VISION 2050, 59

water 8, 11, 15, 19–20, 22, 28, 29, 31–33, 37, 38, 45, 46, 49, 53, 54, 58, 62, 65–76, 82, 91, 94, 95, 97, 98
  as critical climate change issue 67
working class sustainability 46

www.ingramcontent.com/pod-product-compliance
Lightning Source LLC
Chambersburg PA
CBHW030142170426
43199CB00008B/168